We make what matters work.*

ARServices
THE EMPLOYER OF CHOICE FOR VETERANS.
ARServices is the contractor of choice for government organizations and federal teaming partners that require:
Research and development support services
Business transformation services
Integrated logistics services
At ARServices, hiring veterans is essential for our success, and we're proud that veterans and reservists comprise more than 25% of our team.
Find out why veterans choose to work for ARServices at
arslimited.com

It takes Someone Special to Serve your **Country.**

It takes Someone Special to Serve your **Community.**

For your next career, consider healthcare. Consider SENTARA - a nationally recognized health-care system based in the Hampton Roads Virginia area.

It takes a lot to keep one of our nation's top healthcare systems moving, and your skills and military leadership experience are essential ingredients. Whether it's providing direct patient care as a Physician or Nurse, or supporting healthcare delivery in one of the many allied healthcare postitions such as radiology, laboratory, or rehabilitative medicine, your clinical expertise and military background is highly regarded.

Many positions in administration, finance and accounting, IT, and healthcare management are also available with Sentara at our over 300 locations. From Northern Virginia, to North Carolina, and all over the state of Virginia, there are opportunities for your next career in the most stable and personally rewarding environment imaginable. Sentara is a top employer of military families in this area, so when you think about your next career, or even your next job, think of Sentara.

Equal Opportunity Employer
Sentara maintains a drug free workplace

SENTARA®

For employment opportunities, visit
www.sentaracareers.com

Virginia Values Veterans
Official V3 Certified Company

The Virginia Values Veterans (V3) Program is an official Commonwealth of Virginia, Department of Veterans Services Program. © 2014

CAT

Let's Solve Something Important

Thousands of military professionals are bringing their skills and experience to Leidos, supporting vital missions and tackling the world's toughest problems.

Leidos is hosting ongoing monthly virtual chats for transitioning military, veterans, and military spouses. Join us! Please visit the link below for details and to register.

leidos.com/gray

LEIDOS IS SEEKING POWERHOUSE TALENT FOR POSITIONS IN THE AREAS OF IT, ENGINEERING AND SCIENCE

leidos

Preparing professionals for new leadership opportunities is what Maryland Smith does best. Every year, we help hundreds of military veterans and families fearlessly navigate career transitions.
Join our community of Smith Vets and learn how an MBA or MS from one of our top-ranked programs can prepare you for business leadership.
MAKE YOUR
MOVE
EXECUTIVE MBA
FULL-TIME MBA
FLEX MBA (PART-TIME / ONLINE)
SPECIALTY MASTERS
KEEP LEADING FEARLESSLY.
CONTACT US TODAY.
301.405.2559
RHSMITH-VETS@UMD.EDU
RHSMITH.UMD.EDU
UNIVERSITY OF
MARYLAND
ROBERT H. SMITH
SCHOOL OF BUSINESS

BEST FOR VETS 2018 MILITARY TIMES EMPLOYERS
MILITARY FRIENDLY MF'19 BRONZE EMPLOYER
Virginia Values Veterans Official V3 Certified Company
CONTINUE YOUR MISSION
Newport News Shipbuilding, located in Newport News, VA, is the sole designer, builder and refueler of U.S. Navy aircraft carriers and one of two providers of U.S. Navy submarines. We are proud to employ thousands of military veterans and reservists in entry-level and experienced professional and trades careers.
WE ARE HIRING. To apply, visit www.buildyourcareer.com
ENLISTED CAREERS:
ELECTRICAL, ELECTRONICS, HYDRAULICS, MECHANICAL, QUALITY INSPECTORS, TRADES
OFFICER & DEGREED NCO CAREERS:
ENGINEERING, CONSTRUCTION & PROJECT MANAGEMENT, PRODUCTION PLANNERS & SCHEDULERS, SUPPLY CHAIN
Newport News Shipbuilding
A Division of Huntington Ingalls Industries
EEO | Veterans/Disabled Welcome | U.S. Citizenship Required

STOP
Where transitioning service members meet Corporate America
Take 5 minutes. Create your free account now.
www.CorporateGray.com

THE MILITARY TO CIVILIAN TRANSITION GUIDE

SECRETS TO FINDING GREAT JOBS AND EMPLOYERS

CARL S. SAVINO & RONALD L. KRANNICH, PH.D.

www.CorporateGray.com

Corporate Gray

2021 EDITION

3 Great Ways to Get a Job!

Corporate Gray

Connecting employers with military veterans nationwide!

Job Fairs
Online
Books

1. Corporate Gray Job Fairs

Meet with representatives from dozens of military-friendly companies at Corporate Gray's Job Fairs held virtually nationwide or in person in the Washington D.C. metro and Virginia Beach areas. These events are free and especially for job seekers with military experience. Visit www.CorporateGray.com/jobfairs for more information about each event.

Date	Event, Location
Jan 14, 2021	Security Clearance Job Fair, Virtual, Nationwide
Feb 17, 2021	Military-Friendly Job Fair, Virtual, Nationwide
Mar 26, 2021	Military-Friendly Job Fair, Virtual, Nationwide
Apr 23, 2021	Military-Friendly Job Fair, Virtual, Nationwide
May 21, 2021	Military-Friendly Job Fair, Virtual, Nationwide
Jun 25, 2021	Military-Friendly Job Fair, Virtual, Nationwide
Sep 2021	Military-Friendly Job Fair, Virtual or Washington, DC
Oct 2021	Military-Friendly Job Fair, Virtual or Virginia Beach
Nov 2021	Military-Friendly Job Fair, Virtual or Washington, DC

2. Corporate Gray Online

www.CorporateGray.com

Post your resume and search online for jobs of interest among thousands posted by military-friendly companies. Apply directly to those whose openings appear to be a good fit with your skills, interests, and abilities! Get job search advice from the "Transition Guide" and "Blog" tabs.

Subscribers also receive the monthly Corporate Gray eNewsletter with transition and job search advice and hot job opportunities. Don't delay, subscribe today!

3. The Military-to-Civilian Transition Guide

This book is made available to you through the generous support of many military-friendly sponsors. Companies have immediate job opportunities suitable for transitioning military personnel. They welcome receiving your resume and invite you to visit their websites for additional information on their job opportunities. They want to hear from you!

Industry	Company	Website
Defense	ARServices	arslimited.com
	Leidos	Leidos.com/gray
	LMI	lmi.org/careers
	Lockheed Martin	Lockheedmartinjobs.com/military
	Newport News Shipbuilding	www.buildyourcareer.com
	SAIC	saic.com/careers
	Scientific Research Corporation	scires.com
Equipment Dealership	Carter Machinery Company	cartermachinery.com/careers
	Cleveland Brothers Equipment Co.	clevelandbrothers.com/careers
	Foley CAT	Foleyinc.com/careers
Government	Central Intelligence Agency	cia.gov/careers
	National Security Agency	IntelligenceCareers.gov/NSA
	U.S. Postal Service	usps.com/careers
Healthcare	California Correctional Health Care Services	cchcs.ca.gov
	Sentara Healthcare	www.sentaracareers.com
Manufacturing	Caterpillar, Inc.	caterpillar.com
	Eaton Corporation	Eaton.com/MilitaryCareers
School	University of Maryland Smith School of Business	RHSMITH.UMD.EDU

Contents

ISBN-13: 978-0-9996123-4-7

Acknowledgments: Our ability to provide this book to transitioning service members is a direct result of the support of many advertising sponsors. We appreciate the support of these fine organizations and encourage you to take advantage of their transition-related products, services, and career opportunities. You will find information on these sponsors in their recruitment ads. We dedicate this book to you and all those who have served our nation in uniform. May you have a successful career transition!

Quantity discounts are available. For details, please email Susan@CorporateGray.com To order through the GSA Schedule, use GSAadvantage.gov and our Schedule #GS-02F-0146X.

Introduction

Welcome to the Rest of Your Life

LET'S TALK THE TRUTH about your future! Thus far you've had a successful military career. But now you're making a career transition that has major life-changing implications for you and those around you. It's probably unlike any transition you have ever experienced before and unlike any you will ever experience again. With this book you will be able to plan, execute, and manage your transition to the civilian work world.

The purpose of this book is to provide you with the best assistance possible based upon our many combined years of career transition and job search experience. We want to make sure you do this right—clarify your goals, get you moving in the right direction, save you time and effort, minimize your costs, and help you connect with an excellent first post-service job that leads to a new and rewarding career. We've got work to do in the hours, days, and weeks ahead. If you stay with us for the duration of this book, you should acquire many useful tips that will have a significant impact on you and your future worklife.

Question Your Future

All of life's transitions should begin with a series of basic soul-searching questions, for, in posing the right questions, you will begin to develop appropriate responses to your career transition situation.

Let's begin by examining your future and your ability to shape it with these orienting questions:

- What do you want to do with the rest of your life?
- Are you more concerned with finding a rewarding job or moving to a particular location?
- What will you most likely do well and enjoy doing in the civilian workplace?
- How well prepared are you for finding employment in today's job market?
- Do you know how to translate your military experience into civilian terms?

- Are you prepared to research employers, write job-winning cover letters and resumes, network for job leads, conduct job interviews, and negotiate salary and benefits?
- Where do you plan to relocate and what's the job market like in that area?
- If you are married, how involved will your spouse be in your job search?
- Does your spouse share your transition concerns and career goals?

We've got a big, challenging, and exciting task ahead of us. Together we will try to raise the right questions and guide you into answers appropriate to your situation. If you are like many others who have followed the advice of this book, you will discover your career transition work can be both interesting and fun. If done properly, it will make a difference in your life. In fact, it may change your life forever. Along the way, you are likely to renew some old acquaintances and meet new people who will likely have a positive impact on your upcoming life change.

We're delighted to be two of the first new people you're meeting on this road to renewed career success. We're here to help—a good friend and coach if you need one. So let's get started on what should be an exciting journey into a new world of rewarding work.

Life Goes On and On and On

These and numerous other career transition questions and issues provide the central focus for this book. The reality of military jobs and careers is that they eventually come to an end for everyone involved. And they often come to an end faster than most people ever expected. Your time happens to be now. So take a positive, proactive approach to finding the "right" job that leads to a satisfying post-military life.

You're making a career transition in which your past may provide little guidance on how to best chart your future. How you make that transition has important implications for both your personal and professional lives. Therefore, it's extremely important to focus on how you will manage this transition. It's the **transition process**—not different jobs or employers—that should be your central focus. You need a clear sense of where you have been, where you are, and where you want to go with the rest of your life. You do this by focusing on your career transition—past, present, and future.

All good things come to an end, and life does indeed go on. If you've really loved the jobs you had in your military career, chances are you will really love your next series of jobs in your civilian career as well. But you'll have to be diligent at finding your right "fit" in the civilian world. It won't happen overnight and there are no magic pills to make what may be a difficult transition quickly go away. You'll need to do some serious thinking and planning, starting with setting goals, assessing your skills, and charting a course of action aimed at finding the right job for you. You may want to involve other people in this process. If you are married, make sure your spouse is involved early on.

Chances and Choices

Chances are, you are planning to do something rewarding in your next work life. A very small number of transitioning military personnel retire completely. But most military personnel go on to second, third, and fourth careers. Many become employees in large corporations, small businesses, and government. Others decide to start their own businesses. For them, the most critical transition is made upon leaving the service for their first post-military civilian job.

If done properly, this transition can lead to a most rewarding career. If done haphazardly, the transition may result in finding the wrong job, leaving within the first year, and wandering on to other inappropriate jobs. Some people become unhappy job-hoppers who communicate the wrong messages to potential employers. Unfortunately, they repeat this pattern of job disappointment for the rest of their work life. This should not happen to you.

You have excellent skills, experience, and work habits that are readily marketable in the civilian work world. But what you most need to know is how to best present and market your skills and experience. You do this by acquiring another set of skills you may not possess—job search skills. Do you know, for example, how to best write and market a job-winning resume? What networking is, and how to use it in a job search? How to locate the best employers based on your experience and skills? How to use social media in your job search? What the best ways are to get job information, advice, and referrals? How you can negotiate a salary 20 percent higher than expected?

You will be able to clearly answer these job search questions once you develop or refine your job search skills. The chapters that follow help you develop the job search skills for making that all-important career transition.

Best Jobs For the 21st Century

Today, individuals with the right education, skills, and experience are in a better position to find good jobs that should lead to career advancement in the years ahead. In fact, the U.S. Department of Labor's employment statistics to the year 2029 identify the fastest growing occupations (www.bls.gov/ooh/fastest-growing.htm) and the most new jobs (www.bls.gov/ooh/most-new-jobs.htm#). The most new jobs are shown in the chart on the next page. These employment projections confirm what we know is the major trend for the best jobs of the future—they all require greater investments in education and training. For more information on job trends, see the Bureau of Labor Statistics' *Occupational Outlook Handbook* at www.bls.gov/ooh.

If you want a competitive edge in this new global economy, the most important skill you need to acquire is the ability to learn. The best jobs disproportionately favor workers who are constantly learning new skills, adapting to changing markets, and producing goods and services cost-effectively and efficiently. Employers are under pressure to cut production and service costs, and they need workers who are quick to learn and apply new skills.

Getting Started Right

Where do you start, and what should you do first? The process of making a career transition is not really difficult. Almost anyone can find a job. But finding a good job is hard work. Much of what is involved in finding a job is common sense and follows a rational decision-making model as outlined in Chapter 1. In the military, you are used to setting goals, developing plans, and focusing on accomplishing the mission. Therefore, the job-finding process should make good sense to you. You'll initially recognize the process as nothing more than good planning and implementation meeting common sense.

At the same time, you are dealing with a chaotic job market in which information on job vacancies and employers is difficult to access, the rules for finding jobs are inconsistent, and the screening process seems unpredictable and unfair. Rational planning applied to a chaotic environment beyond your control can lead to numerous rejections, disappointments, and frustrations. As you attempt to accomplish a mission in such an uncertain environment, you will discover the wonderful world of **serendipity**—chance occurrences that may unexpectedly lead to the right job for you. Always keep your mind and eyes open for serendipity. It's what makes chaos tolerable, forgiving, and unexpectedly rewarding.

Most New Jobs 2019-2029

Twenty occupations with the highest projected numeric change in employment, sorted by annual median pay.

Occupation	Number of New Jobs (Projected), 2019-29	Annual 2019 Median Pay
Financial managers	108,100	$129,890
Nurse practitioners	110,700	$109,820
Software developers and software quality assurance analysts and testers	316,000	$107,510
Medical and health services managers	133,200	$100,980
General and operations managers	143,800	$100,780
Management analysts	93,800	$85,260
Project management specialists and business operations specialists, all other	79,800	$73,570
Registered nurses	221,900	$73,300
Market research analysts and marketing specialists	130,300	$63,790
Substance abuse, behavioral disorder, and mental health counselors	79,000	$46,240
Medical assistants	139,200	$34,800
Passenger vehicle drivers, except bus drivers, transit and intercity	94,400	$31,340
Landscaping and groundskeeping workers	119,900	$30,440
Nursing assistants	116,900	$29,660
Laborers and freight, stock, and material movers, hand	125,700	$29,510
Cooks, restaurant	327,300	$27,790
Janitors and cleaners, except maids and housekeeping cleaners	105,600	$27,430
Home health and personal care aides	1,159,500	$25,280
Waiters and waitresses	97,600	$22,890
Fast food and counter workers	460,900	$22,740

SOURCE: Occupational Outlook Handbook, U.S. Dept. of Labor, U.S. Bureau of Labor Statistics

Don't Forget Your Concerned Spouse

If you are married, don't forget your spouse on the way to finding the right job. Your spouse shares your concern in finding appropriate civilian employment, and we recommend involving your spouse early in the career transition process. Use each other as a sounding board for exploring career alternatives, examining new ideas, assessing your progress, and keeping your job search focused and realistic. Better still, you may want to conduct two job searches together!

As in most ventures in life, those who persevere succeed. So stay positive, involve your spouse early in the job search process, and enjoy the journey.

Planning and the Art of Sailing Into Your Future

We wish you well as you embark on an exciting yet challenging process of finding a job and career right for you. The process described in Chapter 1 follows nine job search steps involving investigation, written communication, and employer contact. Successful job seekers learn to incorporate these steps into their daily routine.

The chapters that follow are designed to guide you through the key steps in the career transition process. We include the latest job search strategies and techniques that have worked for thousands of job seekers. They will work for you if you organize yourself properly and take the time to put them into practice.

The role of planning in your job search should approximate the art of sailing: you know where you want to go and the general direction for getting there. But the specific path, as well as the time for reaching your destination, will be determined by your environment, situation, and skills. Like the sailor, you tack back and forth, progressing within what is an acceptable time period for successful completion of the task.

Your plan should not become an **end**—it should be a flexible **means** for achieving your stated job and career goals. Planning makes sense, because it requires you to set goals and develop strategies for achieving the goals. However, too much planning can blind you to unexpected occurrences and opportunities. Be sure you stay open to serendipity!

In the end, you'll discover sailing is really what your job search is all about. If you stay with us long enough, you're going to learn to sail very well, perhaps beyond your wildest expectations. You'll learn a great deal about yourself and others in the job market. Best of all, you're going to turn what may initially appear to be a challenge into one of the most exciting times of your life.

We wish you smooth sailing as you chart your course for a new and exciting career and/or lifestyle that will help make a positive transition for the rest of your life.

For more assistance in developing your career transition plan,
visit www.CorporateGray.com > Transition Guide > Develop Your Transition Plan.

chapter 1

Get Organized for New Successes

YOU ARE JOINING MILLIONS of other individuals who go through job and career transitions every year. Indeed, between 15 and 20 million people find themselves unemployed each year. Millions of others try to increase their satisfaction within the workplace as well as advance their careers by looking for alternative jobs and careers.

If you are like most other Americans, you will make more than ten job changes and three to five career changes during your lifetime. You are now engaging in the first of what may become several such changes in your future.

Plan and Organize Your Job Search

Finding a job or changing a career should be done in a systematic and well-planned manner. In this book we will walk you through our 9-step job search process shown in Figure 1-1. The job market can be relatively disorganized and unpredictable, especially compared to the structure you are accustomed to in the military. To be effective your job search must be well-planned, but also flexible. Stay open to opportunities that you may not have previously considered.

Your job search should represent the best of what you are in terms of your past accomplishments as these relate to your future goals. If you base your job search on this concept, you will communicate your best self to employers as well as focus on your strengths both during the job search and on the job.

The individual job search steps are interrelated and most should be followed sequentially. Be sure to properly complete the initial self-assessment steps as this will ensure your job search stays focused. For example, you should never write your resume before conducting an assessment of your skills and identifying your career objective.

Test Your Job Search Competencies

It's important for you to know how to find jobs that are relevant to your background and interests. Before you acquire names, addresses, and phone numbers of potential employers, you should possess the necessary job search knowledge and skills for gathering and using job information effectively.

9 Steps to Job Search Success

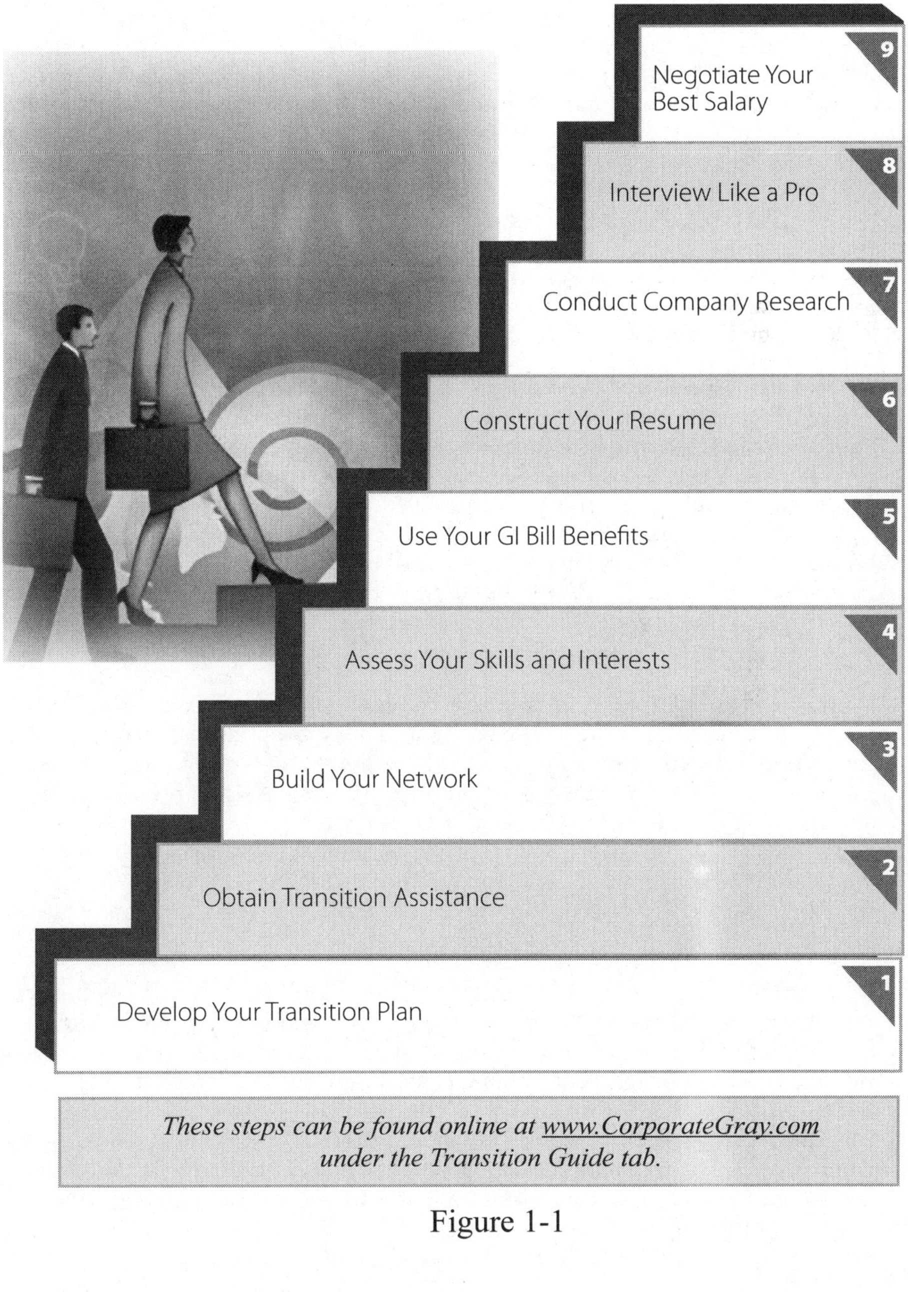

These steps can be found online at www.CorporateGray.com under the Transition Guide tab.

Figure 1-1

Answers to many of your job-related questions are found by examining your present level of job search knowledge and skills. Successful job seekers, for example, use a great deal of information as well as specific skills and strategies for getting the jobs they want.

Let's begin by testing for the level of job search information, skills, and strategies you currently possess as well as those you need to develop and improve. You can easily identify your level of job search competency by downloading and completing The Career Competency Assessment exercise available on the Corporate Gray Online Transition Guide at www.corporategray.com/exercises/career_competency_assessment.pdf.

Using the Career Competency Assessment, you can calculate your overall job search competencies by adding the numbers you circled for a composite score. If your total is more than 75 points, you need to work on developing your job search skills. How you scored each item will indicate to what degree you need to work on improving specific job search skills. If your score is under 50 points, you are well on your way toward job search success. In either case, this book should help you identify job search skills you need to acquire or strengthen.

Career Transition Services

This book augments the transition services offered through your local military transition assistance office. Listed below are descriptions of the various services through which you can receive help in career planning, job search, and employment assistance:

1. **Government-sponsored services:** The Departments of Defense, Labor, and Veterans Affairs offer customized transition services for free. We strongly recommend contacting your nearest transition assistance office as a first step (see Chapter 2). The professional career counselors you encounter will have a wealth of experience and are motivated to assist you in making a smooth transition.

   Your separation or retirement from the military is controlled by your branch of service within the Department of Defense. Within the structure of that program, Department of Labor staff and/or their contracted representatives will conduct many of the transition classes. The Department of Veterans Affairs representative will explain your earned benefits. A more detailed explanation of these programs is in Chapter 2.

2. **Professional associations:** Professional associations often provide placement assistance. If you are not a member, we advise you to join the appropriate association before you leave the service. The primary purpose of these associations is to protect and foster the interests of military personnel—both active and retired. A list of professional military associations can be found in Chapter 3.

   You should also investigate other professional associations, especially those in your desired field of employment. You will find that there is a professional association for almost every civilian occupation. Some associations have job search and placement services for their members. Do an Internet search on "professional associations."

3. **Job fairs:** Many transitioning military have landed jobs through participation in a job fair. This is an excellent way to meet with employers, either in person or virtually, who are interested in hiring people with military skills and experience. Job fairs are typically free and open to job seekers of all ranks (officers and enlisted) and branches of service. For a schedule of the Corporate Gray events, visit www.CorporateGray.com/jobfairs.

   Before attending a job fair you should thoroughly research the participating companies and apply to appropriate positions ahead of time. Be prepared to talk to company representatives about your applications and why you are a match for their needs. Be sure you are "dressed for success," even for a virtual event, which can often include video capability. Make the most of these one-on-one meeting opportunities!

4. **Military placement agencies:** There are many placement agencies in the business of finding military candidates who fit the profiles of their corporate clients. Most placement agencies do not charge the job seeker a fee for their services. Instead, they are paid by the hiring company. Consequently, placement agencies generally only work with those candidates whose skills and experience fit with the needs of their corporate clients.

5. **Temporary employment firms:** Temporary staffing agencies help firms recruit individuals for a wide range of temporary positions as well as full-time employment. Most offer excellent "temp-to-perm" programs which are ideally suited for transitioning military personnel wishing to upgrade their skills and acquire civilian work experience before accepting a full-time position. Indeed, if you are interested in "testing the waters," we recommend that you contact these firms for information on their services. Examples of temporary employment firms include:

    - **Manpower**
    - **Kelly Services**
    - **Adecco**

6. **Testing and assessment centers:** Testing and assessment centers provide assistance for identifying vocational skills and interests. You may wish to use some of these services – in addition to our exercises in Chapter 4 – to target your career and formulate your job search objective. You can find online assessments at the following websites: http://assessment.com and www.self-directed-search.com.

7. **Professional resume writers and career coaches:** There are many reputable professionals that assist job seekers in organizing and implementing an effective job search. Before accessing their services, we strongly recommend you take full advantage of the free career services offered through your local military transition assistance office. If you feel that additional assistance is needed, you can use the following resources to find a certified professional:

    - **Certified Career Coaches** https://thecareerexperts.com/find-career-experts
    - **Professional Association of Resume Writers & Career Coaches** https://parwcc.com
    - **National Board for Certified Counselors, Inc.** www.nbcc.org
    - **National Career Development Association** www.ncda.org

When in Doubt, Take Action

The old adage "When in doubt, do something" is especially relevant when related to the job search process: "When in doubt, engage in a concrete activity related to the sequence of job search steps." This might include conducting research on communities, companies, positions, and salaries; surveying job vacancy announcements; writing a resume and job search letters; or contacting three employers a day.

But developing a plan and taking action is much easier said than done. If conducted properly, a job search can become an extremely time-consuming activity. It inevitably competes with other personal and professional priorities. That's why you need to make some initial decisions as to how and when you will conduct a job search. How much time are you willing to set aside each day or week to engage in your job search activities? After you've spent numerous hours identifying your abilities, skills, and objective, are you willing to commit yourself to 20 hours a week to network for information and advice? Success only comes to those who take action at all stages in the job search process.

Use Your Time Wisely

The amount of time you can devote to your job search will depend on your personal situation. Therefore, as time allows, we recommend you concentrate on your job search sooner rather than later and make maximum use of the time available. There are certain advantages that accrue to seeking a new job while still employed. Because you are currently employed and have key benefits like health coverage, you usually appear more attractive to prospective employers, who may view unemployed candidates as more likely to take the first job available. Remember, your goal is to find an employment opportunity that aligns well with your skills and military experience.

A simple yet effective technique for improving your time management practices is to complete a prioritized "to do" list for each day. Include at the top of your list a particular job search activity or several activities that should be completed that day. If you follow this simple time management practice, you will find the necessary time to include your job search in your daily routine. You can give your job search top priority. Better still, you will accomplish more in less time, and with better results.

Organize Your Job Search

There are certain tools that we suggest using to organize your job search. First, you will want to use a calendar to keep track of your various job search related appointments. Second, establish a task list to help you identify and manage the various tasks that are part of your job search. Use the Planning Worksheet found in Exercises on the Transition Guide page of Corporate Gray Online. Third, use a software package like Excel to help you manage contact information and to track the outcome of your discussions. Fourth, use a filing system with folders and sub-folders to help you find job search items quickly. The more organized you become, the more efficient your job search will be.

Develop an Action-Oriented Timeline

In any mission, it is important to establish a timeline and be disciplined in the execution of your planned activities. It is no less important when embarking on your employment campaign. Your mission is to obtain employment. Therefore, you need to do the following:

1. Determine the number of weeks/months before you need to start the new job.
2. Prioritize the employment search activities.
3. Determine which activities can be done concurrently or must be done sequentially.
4. Place activities on a timeline.
5. Execute the plan.
6. Follow up each job search activity with appropriate actions.

Developing a sound yet flexible job search plan is critical to conducting a successful employment campaign. The time spent planning and organizing your activities will result in a far more effective job search.

A timeline is one technique you should include in your job search for planning your activities. Here's an example of how you may want to sequence your timeline:

FIRST MONTH:

- Develop a job search plan and conduct an assessment of your job search competencies (www.corporategray.com/exercises/career_competency_assessment.pdf).

- Identify your key skills and career interests.
- Contact your local transition assistance office to get information on their services.
- Identify other career transition resources.
- Make a list of your friends and colleagues who could help you with your job search.
- Evaluate your civilian wardrobe and upgrade your business attire as needed.

SECOND MONTH:

- Attend transition classes
- **Prepare your resume:**
  - Study resume examples shown in Chapter 6.
  - Select and download a sample resume from www.corporategray.com/resumes/guest_index.
  - Modify the sample resume as appropriate.
  - Show your resume to a military career counselor and ask for their constructive feedback.
  - Post your resume on Corporate Gray Online (www.CorporateGray.com) and other military affiliated sites.
- **Research employment fields:**
  - Use the Internet (www.LinkedIn.com, www.GlassDoor.com, etc.) to learn about the different companies that exist in your career fields of interest.
  - See our research advice in Chapter 7.
- **Initiate informational interviews:**
  - Contact your friends and associates - build your network (see Chapter 3).
  - Prepare for the interviews—think about questions you will ask; research different fields, companies, and industries.
  - At the interview—ask intelligent questions; listen; thank them for their time.
- Join an association and attend trade shows in your career fields of interest.

THIRD MONTH:

- Continue to expand your network through informational interviews.
- Network with members of associations you've joined.
- Keep track of whom you've met, what transpired, next steps.
- Expand your business wardrobe.
- Identify potential references and contact them; send copies of your resume.
- Continue industry research; identify industry leaders.
- Use Internet employment sites to search and apply for job listings.
- Attend job fairs, both traditional and virtual.

FOURTH AND SUBSEQUENT MONTHS:

- Follow up on all employment leads.
- Stay in touch with your contacts and references.
- Continue to expand your job search network.

Job hunting requires time, careful planning, and hard work, but if done properly can pay off with a job that is right for you.

To help track your progress, download a planning worksheet by visiting www.CorporateGray.com > Transition Guide > Develop Your Transition Plan > Exercises: Planning Worksheet. While there, also download and take the Job Search Competency Assessment.

chapter 2

Obtain Transition Assistance

YOU DON'T HAVE TO FACE your career transition alone. You will find numerous organizations and individuals eager to assist you with career transition and job search services. Many of these services are free, while others can be expensive. The quality of these services varies widely. For the uninitiated, selecting the right services and choosing the best quality organizations and individuals to assist you can present truly bewildering choices. Let's try to sort through this maze.

You need to make the right choices for transition assistance. We recommend starting out "close to home." The Department of Defense, the military services, and various federal and state agencies sponsor high quality transition services you should use throughout your job search. We strongly recommend taking advantage of the seminars, webinars, websites, and other employment services offered by these government agencies. In this chapter we describe these services and discuss how to obtain additional information.

Federal Government Transition Programs and Services

The Department of Defense (DoD), Department of Labor (DoL), and Department of Veterans Affairs (VA) have developed innovative transition resources and programs to assist separating military members, veterans, and their families.

Major Transition Resources and Programs

Department of Defense Transition Assistance Program

The Transition Assistance Program (TAP) provides information and training to prepare military members and their spouses for the transition to civilian life, whether they are pursuing further education, a civilian career, or starting their own business. This interagency collaboration was recently redesigned to improve and standardize the opportunities, services, and training that military members receive across all branches of service. The TAP website at www.dodtap.mil contains resources and information to help you plan and execute your transition.

DoD Military-Civilian Transition Office

The Department of Defense (DoD) Military-Civilian Transition Office (MCTO) has rolled out a new enterprise Learning Management System (LMS) for the Transition Assistance Program (TAP). The new LMS, located at www.tapevents.org/courses, can be accessed any time by transitioning service members, veterans, family members, caregivers and interested members of the public. Curriculum currently available in the LMS include:

- DoL Employment Fundamentals of Career Transition
- DoD Managing Your (MY) Transition
- DoD Military Occupational Codes (MOC) Crosswalk
- DoD Financial Planning for Transition
- VA Benefits & Services

In addition, transitioning military members have the option of participating in one or more tailored tracks aligned with their personal goals, including:

- Accessing Higher Education Training Track
- Career Technical Training Track
- Entrepreneurship Training Track

In-seat learning is the preferred method of delivery for the TAP curriculum. However, many installations have moved to 100% virtual or a hybrid-delivery model for TAP curriculum modules.

Disabled Transition Assistance Program

This program provides servicemembers who are separating or retiring for medical reasons with specialized assistance to learn about VA's Veteran Readiness and Employment Program. Most sites add a half day onto a workshop for these VA-led sessions.

Veterans Benefits Administration

The Department of Veterans Affairs and Department of Defense have created the Veterans Benefits website at www.benefits.va.gov/benefits. This is a central location for veterans, service members, and their families to learn about their benefits. The VA website also has career and employment information and links to employment resources at www.va.gov/careers-employment.

The National Resource Directory (www.nrd.gov) connects service members, veterans, wounded warriors and their caregivers, and family members to programs and support services for recovery, rehabilitation, and community reintegration. Among the many topics of information included are benefits & compensation, education & training, and employment.

Department of Labor Employment Resources

My Next Move for Veterans (www.mynextmove.org/vets) is a DoL-sponsored website to help veterans learn about their career options. Browse careers by keywords or industry, and find careers similar to your military occupation. The O*Net Interest Profiler that can help you find out what you like to do and what kinds of careers you might want to explore. DoL also has a website of employment resources for veterans at www.veterans.gov.

Specialized Transition Resources

Occupational Outlook Handbook

The DoL's Bureau of Labor Statistics (BLS) publishes the Occupational Outlook Handbook (www.bls.gov/ooh) that contains profiles of hundreds of occupations. You can find descriptions of various occupations, work environments, compensation levels, and more. Each profile also includes BLS employment projections for the 2019–2029 decade.

Veteran Readiness and Employment (VR&E)

The VA's Chapter 31 Veteran Readiness and Employment (VR&E) (formerly known as Vocational Rehabilitation and Employment) helps veterans with job training, employment accommodations, resume development, and job seeking skills coaching. Other services may be provided to assist Veterans and Servicemembers in starting their own businesses or independent living services for those who are severely disabled and unable to work in traditional employment. Visit www.benefits.va.gov/vocrehab.

Troops to Teachers

The Troops to Teachers (TTT) Program, operated by the Defense Activity for Non-Traditional Education Support (DANTES), assists eligible military personnel in becoming public school teachers in grades K-12. Visit www.ProudToServeAgain.com.

State Government Employment Programs

State Employment and Training Service Offices

Visit the CareerOneStop website at www.careeronestop.org for information on careers, salary, education and training, job search, and state resources. Find local help by searching for American Job Centers in your area at www.careeronestop.org/LocalHelp/AmericanJobCenters/find-american-job-centers.aspx.

Veterans receive special consideration and priority for referral, testing, and counseling from state employment offices, which provide many additional services such as training opportunities, information on state employment and apprenticeship programs, and statistics regarding employment availability, economic climate, and cost-of-living data.

LVERs and DVOPs

Within each state's employment services office, you will find Local Veterans' Employment Representatives (LVERs) and Disabled Veterans' Outreach Program (DVOP) specialists who are trained to help you find local job opportunities. They monitor and oversee veterans' employment services, administer veterans' training programs under the Job Training Partnership Act, and protect the reemployment rights of veterans. These representatives can assist veterans with any employment problem. Visit your local American Job Center to connect with the LVERs and DVOPs in your area. Also visit CareerOneStop's Job Search online resources (https://www.careeronestop.org/JobSearch/job-search.aspx) for more online employment assistance.

Military Transition Services

Army Programs

Soldier for Life

The Army's Soldier For Life - Transition Assistance Program (SFL-TAP), formerly called the Army Career and Alumni Program (ACAP), is the principal program the Army offers to assist you in transitioning to your next career. Its mission is to provide you with timely and effective transition assistance. To accomplish this, centers are operated at major installations around the world. Visit www.sfl-tap.army.mil.

You are required to attend a pre-separation briefing no less than 90 days prior to your leaving the Army. During this briefing, you will learn about your career options, transition benefits, and available services. In addition, you will be tested on your career transition readiness and directed down the appropriate path.

Soldier and Family Assistance Centers (SFACs)

The Army's Soldier and Family Assistance Centers (SFACs) are close to the Army Wounded Warrior Battalions. The SFAC includes the services of a transition counselor who provides Army wounded warriors with one-on-one career counseling and job placement assistance.

ACS Employment Readiness Program (ERP)

The Employment Readiness Program (ERP), operated by the Army Community Service (ACS), provides these services to adult military family members:

- Resume preparation
- Interviewing techniques
- Career counseling
- Job listings
- Job referrals
- Federal application preparation

Navy/Marine Programs

Fleet and Family Support Center / Community Services

The Navy's Fleet and Family Support Centers (FFSC) support individual and family readiness through programs and resources that help Navy families be well-informed and adaptable to the Navy environment. Programs and services are offered at Fleet and Family Support Centers worldwide. For more information, visit the CNIC Headquarters website at www.cnic.navy.mil.

The Marine Corps' Community Services offer many support programs, including transition assistance. Visit their website at www.usmc-mccs.org.

Air Force Programs

Military and Family Support Centers

The Air Force Military and Family Support Centers (MFSC) support the total force, (active duty, retired military, DOD civilians, Guard and Reserve) and their eligible family members. The MFSC also helps commanders and base leaders with the health and welfare of military families and single members.

The MFSC offers group classes and individual appointments in several different areas, including budget or finances, PCS moves, volunteering, employment assistance, Air Force Aid, or information on deployment programs and benefits. By participating in the many programs offered through the MFSC, you can make the most of your career change. Contact your local MFSC for further information.

Coast Guard Programs

Coast Guard Transition/Relocation Managers

The cornerstone of the U.S. Coast Guard's transition program are the Transition/Relocation Managers located within the Work-Life Staff. The Transition/Relocation Managers conduct both the Retirement and Separation Seminars and assist members on an individual basis. Ideally, separating members should attend the program seminars a minimum of six months before the end of their commitment.

Anyone wishing to attend a seminar should contact the Transition/Relocation Manager located in their regional Work-Life Staff. For assistance, visit www.uscg.mil.

Military Spouse Programs

Spouse Employment Assistance Program (SEAP)

The Spouse Employment Assistance Program (SEAP) assists spouses overcoming the difficulties associated with finding employment, especially during the relocation process.

If you wish to use SEAP services, you should contact the SEAP office at the Fleet and Family Support Center nearest you. If you are in the Coast Guard, contact the Transition/Relocation Manager located at the Work-Life Staff nearest you.

Military Spouse Employment Partnership (MSEP)

The MSEP website (https://msepjobs.militaryonesource.mil) offers military spouses job search assistance and connects them with employers seeking their skills.

To view a compiled list of free career transition services, visit www.CorporateGray.com > Transition Guide > Obtain Transition Assistance.

chapter 3

Network Your Way to Career Success

NETWORKING IS ESSENTIAL to a successful job search, and it should be started early in the job search process. It is estimated that 80% of jobs are found through networking. Growing your network and connecting to those who can help will increase the opportunities for job leads, advice, and support in your job search. Most people send out cover letters and resumes in response to job listings, then wait to be called for an interview. This "direct-mail" passive approach will yield a disappointing response rate. Being proactive by using interpersonal and networking strategies in which the resume plays a supportive rather than primary role will yield better success. With this approach you develop an informal information network you will use to gather knowledge about available jobs. These contacts and information become key building blocks for generating job interviews and offers.

There are millions of jobs that go unadvertised. To identify those opportunities, you'll need to crack the hidden job market. Research and experience show that the most effective means of communication are face-to-face and word-of-mouth. The informal, interpersonal system of communication is the central nervous system of the hidden job market. Methods for making important job contacts are **networking, prospecting**, and using **informational interviews** to get useful job information.

Communicate Your Qualifications

Taken together, these interpersonal methods help you **communicate your qualifications** to employers. Although many job seekers may be reluctant to use this informal communication system, they greatly limit their potential for success if they do not.

Put yourself in the position of the employer for a moment. You have a job vacancy to fill. If you advertise the position, you may be bombarded with hundreds of resumes, applications, phone calls, etc. While you do want to hire the best qualified individual for the job, you simply don't have time to review scores of applications. Furthermore, with limited information from application forms, cover letters, and resumes, you find it hard to identify the best qualified individuals to invite for an interview.

So, as an employer, what do you do? You begin by calling your friends, acquaintances, and other business associates and ask if they or someone else might know of any good candidates for the position. If they can't help, you ask them to give you a call should they learn of anyone qualified for your vacancy. You, in effect, create your own hidden job market—an informal information network for locating desirable candidates. Your trusted contacts initially screen the candidates in the process of referring them to you. This both saves you time and minimizes your risks in hiring a stranger.

Based on this understanding of the employer's perspective, what should you do to best improve your chances of getting an interview and job offer? Networking for information, advice, and referrals should play a central role in your overall job search. Remember, employers need to solve personnel problems. By conducting informational interviews and networking, you help employers identify their needs, limit their alternatives, and thus make decisions and save money. At the same time, you gain several advantages by conducting these interviews:

1. You are less likely to encounter rejections since you are not asking for a job—only information, advice, and referrals.
2. You go after unadvertised positions.
3. You encounter less competition.
4. You go directly to the people who have the power to hire.
5. You are likely to be invited to interviews based upon the referrals you receive.

Most employers want more information on candidates to supplement the "paper qualifications" represented in application forms, resumes, and letters. Studies show that employers in general seek candidates who have these skills: communication, problem solving, analytical assessment, and planning. Surprising to many job seekers, technical expertise ranks third or fourth on employers' lists of most desired skills. Therefore, communicating your qualifications to employers entails more than just informing them of your technical competence. You must communicate that you have the requisite personal and professional skills for performing the job. Informal prospecting, networking, and informational interviewing activities are the best methods for communicating your qualifications to employers.

Develop Networks

Networking is the process of purposefully developing relations with others. Networking in the job search involves connecting and interacting with other individuals who can be helpful to you. Your network consists of everyone with whom you interact. The more you develop, maintain, and expand your networks, the more successful your job search should be.

Your basic network encompasses the following: family, friends, acquaintances, professional colleagues, supervisors, and local business professionals. As a member of the military, you already have an extensive network in place. Based on your many moves and the scores of people you have come in contact with through your military service, you should be well positioned to take advantage of this important employment avenue. Identify everyone in your network who might help you with your transition, and contact these individuals for advice relating to your job search. It is time to begin networking your way to career transition success.

Your most important network may be your military contacts—both those still in the service and those who have separated or retired. Let your military colleagues know that you are actively searching for a job. You will be pleasantly surprised by the positive response you will receive. Indeed, the military tends to be a close knit family that looks out for each other. Military friends and associates who have already transitioned to the private sector can be especially helpful, for they know first-hand where the hidden jobs are within their organizations. You might want to begin by referring to your holiday card list. You can use the following websites to contact your former military colleagues: www.vetfriends.com and www.linkedin.com.

Join Military Associations

As a transitioning service member, you have the opportunity to join a variety of military associations, many of which offer job search assistance. Being a member of an association significantly expands your opportunity to network with other veterans, many of whom are employed in the private sector. We encourage you to join one or more of these associations:

- Military Officers Association of America www.moaa.org
- Non Commissioned Officers Association www.ncoausa.org
- Reserve Officers Association www.roa.org
- The Air Force Association www.afa.org
- Association of the United States Army www.ausa.org
- U.S. Army Warrant Officers Association www.usawoa.org
- Navy League of the United States www.NavyLeague.org
- Fleet Reserve Association www.fra.org
- Marine Executive Association www.marineea.org
- Marine Corps Reserve Association www.usmcra.org
- Marine Corps League www.mclnational.org
- Marine for Life usmc-mccs.org/services/career/marine-for-life-network
- Armed Forces Communications and Electronics Association www.afcea.org
- American Legion www.legion.org
- Veterans of Foreign Wars www.vfw.org
- Disabled American Veterans www.dav.org
- AMVETS amvets.org

Use Social Media

Be sure to effectively use social media tools such as LinkedIn, Twitter, and Facebook to extend your network. In so doing, you will maximize the chances of your friends—and people you may not yet know but share a common interest—helping you identify job leads.

LinkedIn (www.linkedin.com) is a professional networking tool that helps you stay in contact with past and present colleagues, find job opportunities, identify key individuals within companies of interest, and get introduced to those who can help you get the job. If you have not yet created a LinkedIn account, we encourage you to do so. Fully complete your profile using keywords that describe your skills, and give special attention to the Headline. By making it enticing, you make it more likely that someone—like a recruiter—will want to contact you.

LinkedIn lets you share updates, and we recommend you post your status so that your connections stay apprised of your job search progress. LinkedIn supports Groups to help you build a relevant network. Join and participate in military-affiliated groups, especially those for military-to-civilian career transition, as well as groups in your industry. Also take advantage of LinkedIn's current offer of a free premium account upgrade for military veterans. Visit LinkedIn's veterans page to learn more (www.veterans.linkedin.com).

Twitter (www.twitter.com) allows you to network with people you may not know but share a common interest. Twitter posts, or tweets, are short, text-based comments. Twitter moves quickly and can be a good source of job search information if you keep up with it.

By tweeting, you can let your followers know about your job search. You can also tweet about your industry to help establish your online presence. Use keywords (your most relevant skills) when filling out your Twitter bio.

Use Twitter to follow experts in your industry, career coaches, companies, and recruiters to learn about job openings at companies and gain insight into those jobs. There are Twitter accounts set up just for job listings, so if you follow them you'll get their job posts to your timeline as they are tweeted. Be proactive and "follow" as many people and company accounts as appropriate to get fast information about job opportunities.

Facebook (www.facebook.com) is a tool for primarily connecting with friends and reconnecting with people from your past. It, too, can be an effective networking tool. For example, you might post your status on Facebook letting people know that you are leaving the military and seeking civilian employment. By frequently updating your status, you are enabling your friends on Facebook to stay current with what's happening in your life and in your job search. Facebook has a "Jobs" section that adds job search functionality to your account.

When using social media sites such as Facebook, keep privacy in mind. If your Facebook account is used in too casual a manner to be appropriate for a recruiter, we recommend you change your privacy settings accordingly. Whereas sites such as LinkedIn are used only professionally and can be kept public.

The Corporate Gray Blog (http://blog.corporategray.com) has articles on using social media in your job search. Corporate Gray's book, *Social Media and Your Job Search: Maximizing Your Network for a Successful Transition* is also a helpful resource for using social media in your job search. Given the wide expanse of social media, you now have access to a global network of potential contacts at minimal to no cost. Take advantage of these online networking resources!

Prospect For Leads

The key to successful networking is an active and routine prospecting campaign. Salespersons in insurance, real estate, and other direct-sales businesses understand the importance and principles of prospecting. The basic operating principle is probability: the number of sales you make is a direct function of the amount of effort you put into developing new contacts and following through. Expect no more than a 10-percent acceptance rate: you need to contact 100 people for 10 successes. The more people you contact, the more acceptances you will receive.

These prospecting principles are extremely useful for making a career change. Like sales situations, the job search is a highly ego-involved activity often characterized by numerous rejections accompanied by a few acceptances. Don't take a "no" as a sign of personal failure. If you persist long enough, you will achieve success after a few more "no's." Furthermore, if your prospecting activities are focused on gathering information rather than making the "sale," you will considerably minimize the number of rejections. Therefore, you should do the following:

- Prospect for job leads.
- Accept rejections as part of the game.
- Link prospecting to informational interviewing.
- Continue prospecting for more information and "yes" responses, which will eventually translate into job interviews and offers.

A good prospecting pace as you start your search is to make two new contacts each day. Start by contacting people in your immediate network. Let them know you are conducting a job search, but emphasize that you are only doing research. Ask for a few moments of their time to discuss your information needs. You are only seeking information and advice at this time—not a job.

It should take you only a few minutes to make a contact by email, phone, or via social media. If you make two to four contacts each day, by the end of the week you will have 10 to 20 new contacts. The more contacts you make, the more useful information, advice, and job leads you will receive. If your job search slows down, you probably need to increase your prospecting activities. Expect each contact to refer you to two or three others who will also refer you to others.

Initiate Informational Interviews

The most important networking method is the informational interview. Based on prospecting and networking techniques, these interviews minimize rejections and competition as well as quickly open the doors to organizations and employers. If you want a job interview, you first need to understand the informational interview and how to initiate and use it effectively.

Practice the 5 Rs of Informational Interviewing

The guiding principle behind prospecting, networking, and informational interviews is this: **the best way to get a job is to ask for job information, advice, and referrals; never ask for a job.** Remember, you want your prospects to engage in the 5 Rs of informational interviewing:

- **Reveal** useful information and advice.
- **Refer** you to others.
- **Read** your resume.
- **Revise** your resume.
- **Remember** you for future reference.

If you follow this principle, you should join the ranks of many successful job seekers.

Approach Key People

Whom should you contact within an organization for an informational interview? Contact people who are busy, who have the power to hire, and who are knowledgeable about the organization.

You should use a more formal approach to gain access to referrals and new contacts. The best way to initiate a contact with a prospective employer is to approach them on Linkedin. The elements for sending an approach email or message include:

1. **OPENING:** If you have a referral, tell the individual you are considering a career in ________. His or her name was given to you by ________ who suggested he or she might be a good person to give you useful information about careers in ________. Should you lack a referral to the individual and thus must use a "cold call" approach to making this contact, you might begin your letter by stating that you are aware he or she has been at the forefront of ________ business—or whatever is both truthful and appropriate for the situation. A subtle form of flattery will be helpful at this stage.

2. **REQUEST:** Demonstrate your thoughtfulness and courtesy by mentioning that you know he or she is busy. You hope to schedule a mutually convenient time for a brief meeting to discuss your questions and career plans. Most people will be flattered by such a request and happy to talk with you about their work.

3. **CLOSING:** In closing the correspondence, mention that you will call the person to see if an appointment can be arranged. Be specific by stating the time and day you will call—for example, Thursday at 2 pm. **You** must take initiative to follow up the letter with a definite contact time. Use Microsoft Outlook's calendar feature to lock-in the day and time.

4. **ENCLOSURE:** Do NOT enclose your resume with this approach letter. You should take your resume to the interview and present it as a topic of discussion near the end of your meeting. If you send it with the approach letter, you communicate a mixed and contradictory message. Remember your purpose for this interview: to gather information and advice, NOT ask for a job.

Many people will meet with you, assuming you are sincere in your approach. However, many people also are busy and simply won't have the time to meet. If the person puts you off when you phone them for an appointment, clearly state your purpose and emphasize that you are not looking for a job with this person—only information and advice. Make the best of the situation and try to conduct the informational interview over the phone. Send a follow-up email in which you again state your intended purpose and ask to be remembered for future reference. Attach your resume to this email.

While you are ostensibly seeking information and advice, treat these meetings as important preliminary interviews. You need to communicate your qualifications—that you are competent, intelligent, honest, and likable, just as you would in a formal job interview. Follow the same advice given for conducting a formal interview, such as dressing appropriately for a face-to-face meeting (even if virtual).

Conduct the Informational Interview

An informational interview will be relatively unstructured. Since you want the interviewer to give you advice, you reverse roles and you become the interviewer. You should structure this interview with a particular sequence of questions. Most questions should be open-ended, requiring the individual to give specific answers based upon his or her experiences.

Begin your interview by stating your appreciation for the individual's time:

> *"I want to thank you again for scheduling this meeting. I know you're busy, and I appreciate you making time to see me on a subject which is important to my future."*

Next, you should reiterate your purpose as stated in your letter:

> *"As you know, I am exploring job and career alternatives. I know what I do well and what I want to do. But before I commit myself to a new job, I need to know more about various career options. I thought you would be able to provide me with some insights into career opportunities, job requirements, and possible problems or promising directions in the field of ____________."*

This statement normally will get a positive reaction from the person, who may want to know more about what you want to do. Be sure to clearly communicate your job objective.

Your next line of questioning should focus on "how" and "what" questions centering on (1) specific jobs and (2) the job search process. Begin by asking about specific job aspects:

- Duties and responsibilities
- Knowledge, skills, and abilities required
- Work environment relating to fellow employees, work flows, deadlines, stress, initiative
- Advancement opportunities and outlook
- Salary ranges

Your informer will probably take considerable time talking about his or her experience in each area. Listen carefully to what they say.

Your next line of questioning should focus on your job search activities. You need as much information as possible on how to:

- Acquire the necessary skills
- Best find a job in this field
- Overcome any objections employers may have to you
- Uncover job vacancies which may not be advertised and develop job leads
- Approach prospective employers

Your final line of questioning should focus on your resume. Do not show your resume until you address this last set of questions:

- Is this an appropriate type of resume for the jobs I have outlined?
- If an employer received this resume, how do you think he or she would react to it?
- What do you see as possible weaknesses or areas that need to be improved?
- What should I do with this resume?
- How might I improve the form and content of the resume?

You should receive useful advice on how to strengthen both the content and use of your resume. Most important, these questions force the individual to read your resume which, in turn, may be remembered for future reference.

Your last question is especially important in this interview. You want to be both remembered and referred. Some variation of the following question should help:

> *"I really appreciate your advice. It is very helpful and it will improve my job search considerably. Could I ask you one more favor? Do you know two or three other people who could help me with my job search?*

Before you leave, mention one more important item:

> *"During the next few months, should you hear of any job opportunities for someone with my qualifications, please feel free to pass my name on to them."*

Send a thank-you letter or email after completing this informational interview. Express your gratitude, reiterate your interests, and ask to be remembered and referred to others.

Follow up on any useful advice you receive, particularly referrals. Approach referrals in the same manner you approached the person who gave you the referral.

If you continue prospecting, networking, and conducting informational interviews, you will soon be busy conducting formal job interviews and receiving job offers.

For more assistance in building your job search network and for networking etiquette and tips, visit www.CorporateGray.com > Transition Guide > Build Your Network.

chapter 4

Identify Your Skills, Interests, and Values

WHAT SKILLS DO YOU POSSESS that are most relevant to today's job market? Are the skills you acquired in the military attractive to many employers? What other skills do you possess which may or may not be related to your work in the military? Do you need to acquire new skills?

We live in a skills-based society where individuals market their skills to employers in exchange for money, position, and power. The ease by which individuals change jobs and careers is directly related to their ability to communicate their skills to employers and then transfer their skills to new work settings.

To best position yourself in the job markets of today and tomorrow, you should pay particular attention to refining your present skills as well as achieving new and more marketable skills. This may mean going back to school for a degree or certificate or taking advantage of employer-sponsored training programs.

Types of Skills

Most people possess two types of skills that define their accomplishments and strengths as well as enable them to enter and advance within the job market: work-content skills and functional skills. You need to acquaint yourself with these skills before communicating them to employers.

Work-Content Skills

We assume you have already acquired certain work-content skills necessary to function effectively in today's job market. These "hard skills" are easy to recognize since they are often identified as "qualifications" for specific jobs. They are the subject of most educational and training programs. Work-content skills tend to be technical and job-specific in nature. Examples of such skills include repairing helicopters, developing software, teaching history, or operating an X-ray machine. They may require formal training, have an association with specific trades or professions, and be used only in certain job and career settings. One uses a separate skills vocabulary, jargon, and subject matter for specifying technical qualifications of individuals entering and advancing in an occupation. While these skills do not transfer well from one occupation to another, they are critical for entering and advancing within certain occupations.

Functional/Transferable Skills

You also possess numerous functional/transferable skills that employers seek along with your work-content skills. These "soft" skills, which are associated with numerous job settings, are mainly acquired through experience rather than formal training and can be communicated through a general vocabulary. Functional/transferable skills are harder to recognize since they tend to be linked to certain personal characteristics (energetic, intelligent, likable) and the ability to deal with processes (communicate, solve problems, motivate) rather than do things (build a house, repair air conditioners). While most people have only a few work-content skills, they may have numerous functional/transferable skills. These skills enable job seekers to more easily change jobs.

In contrast to work-content skills, functional skills can be transferred from one job or career to another. Functional skills enable individuals to make some job and career changes without acquiring additional education and training. These skills constitute an important bridge for moving from one occupation to another. Examples of skills that are directly transferable to business and industry include leadership, discipline, and critical thinking.

Before you decide if you need more education or training, you should first assess both your functional and work-content skills to see how they can be transferred to other jobs and occupations. Once you do this, you should be better prepared to communicate your qualifications to employers.

Identify Your Skills

Skills Exercises

Use the exercises in this chapter to identify both your work-content and transferable skills. These self-assessment techniques stress your strengths and should generate a rich vocabulary for communicating your qualifications to employers.

To assist you in identifying your **work-content skills**, collect all the efficiency/performance reports you received while serving in the military. As you read through these reports, think about the responsibilities you held and your accomplishments in each. Use your efficiency reports to jog your memory, and ask yourself these questions:

- Did you hold any leadership positions? If so, how many people reported to you?
- Did you have a budget and, if so, how much was it?
- What were the significant skills you developed during that period?
- What were your important accomplishments? Can you quantify them?
- Did you qualify to hold a government security clearance?
- Did you do any volunteer work that might show how you reached out to help others?
- Were you ever stationed overseas? If so, did you learn to speak a foreign language?
- Are you already bilingual?
- Did you work on a college or advanced degree?

You can use the Work-Content Skills List form found on the Corporate Gray website (www.CorporateGray.com) to record your skills (see the Exercises in Step 4 of the Transition Guide tab).

Most **functional/transferable skills** can be classified into two general skills and trait categories—organizational/interpersonal skills and personality/work style traits. Use the Functional Skills Checklist in Figure 4-1 to identify your skills within the two categories. Place a "1" in front of the skills that strongly characterize you, a "2" next to those that describe you to a large extent, and a "3" before those that describe you to some extent.

After completing this exercise, review the list and rank order the 10 characteristics that best describe you on each list. This skills vocabulary helps you better identify and translate your military work experience into civilian occupational language.

Your Strengths

Regardless of what combination of work-content and functional skills you possess, a job search must begin with identifying your strengths. Without knowing these, your job search will lack content and focus. After all, your goal should be to find a job that is a fit for you rather than one you think you might be able to fit into. Of course, you also want to find a job for which there is demand. This particular focus requires a well-defined approach to identifying and communicating your skills to others. You can best do this by asking the right questions about your strengths and then conducting a systematic self-assessment of what you do best. The counselors in the career transition offices are skilled in helping you identify the functional and work-content skills you possess. We caution you not to sell yourself short. Regardless of which branch you served, we know that you gained experience in many areas. Now your challenge is to identify those skills by asking the right questions.

Military-to-Civilian Skills

What skills did you acquire in the military that are directly transferable to the civilian work-world? Employers want to know about your skills and experience. You should be able to explain what it is you do well and give examples relevant to employers' needs. A good place to begin identifying your transferable skills is VMET, the Verification of Military Experience and Training documents.

Verification of Military Experience and Training (VMET)

The Verification of Military Experience and Training (DD Form 2586) is provided to all eligible departees to verify their military experience, training history, associated civilian equivalent job title(s), and recommended educational credit information. This helps service members verify previous experience and training for potential employers, write their resumes, interview for jobs, negotiate credits at schools, and obtain certificates or licenses.

Military-to-Civilian Skills Translators

The Military Occupational Classification (MOC) crosswalk has been updated by the Department of Defense and incorporated into O*NET Online (www.onetonline.org/crosswalk/MOC) and My Next Move for Veterans (www.mynextmove.org/vets). Over 10,000 MOC occupations are linked to related O*NET-SOC occupations. The Career One Stop website also has a skills translator at www.careeronestop.org/Toolkit/Jobs/match-veteran-jobs.aspx, and the VA offers a CareerScope assessment at www.va.gov/careers-employment/careerscope-skills-assessment.

If you are unsure as to which civilian occupations are best suited for the skills you developed in the military, these are good websites to use for research. The military-to-civilian skills translators can also help you put your military skills into civilian language that recruiters will understand.

Functional Skills Checklist (Figure 4-1)

Organizational and Interpersonal Skills

__ communicating
__ problem solving
__ analyzing/assessing
__ planning
__ decision-making
__ innovating
__ thinking logically
__ evaluating
__ identifying problems
__ synthesizing
__ forecasting
__ tolerating ambiguity
__ motivating
__ leading
__ selling
__ performing
__ reviewing
__ attaining
__ team building
__ updating
__ coaching
__ supervising
__ estimating
__ negotiating
__ administering
__ trouble-shooting
__ implementing
__ self-understanding
__ understanding
__ setting goals
__ conceptualizing
__ generalizing
__ managing time
__ creating
__ judging
__ controlling
__ organizing
__ persuading
__ encouraging
__ improving
__ designing
__ consulting
__ teaching
__ cultivating
__ advising
__ training
__ interpreting
__ achieving
__ reporting
__ managing

Personality and Work Style Traits

__ diligent
__ patient
__ innovative
__ persistent
__ tactful
__ loyal
__ risk taker
__ versatile
__ enthusiastic
__ outgoing
__ expressive
__ adaptable
__ democratic
__ resourceful
__ determined
__ creative
__ open
__ objective
__ warm
__ orderly
__ tolerant
__ frank
__ cooperative
__ dynamic
__ self-starter
__ precise
__ sophisticated
__ honest
__ reliable
__ perceptive
__ assertive
__ sensitive
__ astute
__ effective
__ easygoing
__ calm
__ flexible
__ efficient
__ punctual
__ receptive
__ diplomatic
__ self-confident
__ tenacious
__ discrete
__ talented
__ empathetic
__ tidy
__ candid
__ adventuresome
__ firm
__ sincere
__ initiator
__ diplomatic

Civilian Credentialing

You have gained valuable experience through your military service. Now may be a good time to take stock of the skills you have acquired and what new skills you may need to pursue a civilian career. Depending on your circumstances, it may make sense for you to pursue a certificate or earn civilian credentials in your field.

To assist you in identifying civilian credentialing requirements, visit Credentialing Opportunities On-Line (COOL) for the Army, Navy, Air Force, Marines and Coast Guard at www.cool.osd.mil. Also, check out the CareerOneStop Credentials Center (www.careeronestop.org/Credentials) and the credentialing sections of GoArmyEd (www.goarmyed.com), and DANTES (www.dantes.doded.mil).

For those of you who decide that now is a good time to obtain your bachelor's or master's degree, you'll have to decide whether you want to do it on a part-time or full-time basis. If you do make the investment in time now to pursue this degree, how will it increase your income potential 5, 15, or 25 years from now? Are you able financially to go to school full time or would it make more sense to find full-time employment immediately and then pursue your degree on a part-time basis? If you have just completed an educational program relevant to today's job market, the skills you have to offer are most likely related to the subject matter you studied. As you transition from the military, the skills you wish to communicate to employers will most likely be those you have already demonstrated in specific military jobs. If your degree or certificate is in the same area, you can use it as a leverage to increase your market value to a prospective employer. If, on the other hand, the degree or certificate is in a totally different area, you can show that you have received academic training in this particular discipline. Depending on your circumstances, it may be worthwhile to find temporary employment or volunteer work in that area, so that you can later refer to both your training and work experience.

> Now would be a good time to take stock of the skills you have and the skills you need.

The skills required for finding a job are no substitute for the skills necessary for doing the job. Learning new skills requires a major investment of time and effort. Nonetheless, the long-term pay-off should more than justify the initial costs. Indeed, research shows that well selected education and training provide the best returns on your investment. Be sure you understand and take advantage of the excellent educational benefits you earned through your military service. Many who have traveled the road before you know that education is a key to success. We encourage you to unlock the secrets of your potential.

Interests and Values

Knowing what you do well is essential for understanding your strengths and for linking your capabilities to specific jobs. However, just knowing your abilities and skills will not give your job search the direction it needs for finding the right job. You also need to know your work values and interests. These are the basic building blocks for setting goals and targeting your abilities toward certain jobs and careers.

Take, for example, the Administrative Specialist who types 90 words a minute. While this person possesses a marketable skill, if they don't enjoy using this skill and are more interested in working outdoors, this will not become a motivated skill; the individual will most likely not pursue an administrative position. Your interests and values will determine whether or not certain skills should play a central role in your job search.

Vocational Interests

We all have interests. Most change over time. Many of your interests may center on your present job, whereas others relate to activities that define your hobbies and leisure activities. A good place to start identifying your interests is by examining the information and exercises found in *The Complete Guide For Occupational Exploration*. Widely used by students and others first entering the civilian job market, it is also relevant to individuals who already have work experience. The guide classifies all jobs in the U.S. into 12 interest areas. Examine the list of interest areas in Figure 4-2. In the first column, check those work areas that appeal to you. In the second column, rank order those areas you checked in the first column. Start with "1" to indicate the most interesting:

My Work Interests (Figure 4-2)

Organizational and Interpersonal Skills

Yes (✓)	Ranking (1, 2, 3...)	Interest Area
____	____	**Artistic:** an interest in creative expression of feelings or ideas.
____	____	**Scientific:** an interest in discovering, collecting, and analyzing information about the natural world, and in applying scientific research findings to problems in medicine, the life sciences, and the natural sciences.
____	____	**Plants and animals:** an interest in working with plants and animals, usually outdoors.
____	____	**Protective:** an interest in using authority to protect people and property.
____	____	**Mechanical:** an interest in applying mechanical principles to practical situations by using machines or hand tools.
____	____	**Industrial:** an interest in repetitive, concrete, organized activities done in a factory setting.
____	____	**Business detail:** an interest in organized, clearly defined activities requiring accuracy and attention to details.
____	____	**Selling:** an interest in bringing others to a particular point of view by personal persuasion, using sales and promotional techniques.
____	____	**Accommodating:** an interest in catering to the wishes and needs of others, usually on a one-to-one basis.
____	____	**Humanitarian:** an interest in helping others with their mental, spiritual, social, physical, or vocational needs.
____	____	**Leading and influencing:** an interest in leading and influencing others by using high-level verbal or numerical abilities.

Computerized and Online Assessment Systems

While the previous self-directed exercises required you to either respond to checklists of skills or reconstruct and analyze your past job experiences, there are many computerized self-assessment programs designed to help individuals identify their skills.

Many of these programs, like Career Information System (CIS), Guidance Information System (GIS), and SIGI3 (System of Integrated Guidance and Information), are available through schools, colleges, and libraries. You could check with the career or counseling center at your local community college to see what computerized career assessment systems are available for your use. They are relatively easy to use and they generate a great deal of useful career planning information. Many will print out an analysis of how your interests and skills are related to specific jobs and careers.

Also, your career transition assistance office will likely offer interest assessment tests, including the Strong Interest Inventory and the Myers-Briggs Type Indicator. Keep in mind that not all testing and assessment instruments used by career counselors are equally valid for career planning purposes. While the Strong Interest Inventory appears to be the most relevant for career decision-making, the Myers-Briggs Type Indicator is also popular.

In addition to professionally administered and interpreted tests, there is also Holland's "The Self-Directed Search," found at www.self-directed-search.com. The O*Net Interest Profiler (www.mynextmove.org/explore/ip) can help you determine your interests and how they relate to the world of work, as well as the VA's CareerScope assessment (www.va.gov/careers-employment/careerscope-skills-assessment).

Work Values

Work values are those attributes you developed on the job; for example, loyalty, dedication, team playing, and ability to work under pressure. Most jobs involve a combination of likes and dislikes. Figure 4-3 can help you identify your work values. Put a check next to each attribute that most satisfies you about work, and put an "X" next to those you dislike.

My Work Values (Figure 4-3)

___	contribute to society	___	be creative
___	have contact with people	___	supervise others
___	work alone	___	work with details
___	work with a team	___	gain recognition
___	compete with others	___	acquire security
___	make decisions	___	make money
___	work under pressure	___	help others
___	use power and authority	___	solve problems
___	acquire new knowledge	___	take risks
___	be a recognized expert	___	work at my own pace

Select four work values from the above list which are the most important to you and list them below. List any other work values (desired satisfactions) which were not listed above but are nonetheless important to you:

1. ______________________________
2. ______________________________
3. ______________________________
4. ______________________________

____

By identifying what you both like and dislike about jobs, you are better able to identify jobs that involve tasks you will most enjoy.

To further assist you in assessing your skills and interests,
visit www.CorporateGray.com > Transition Guide > Assess Your Skills & Interests.

chapter 5

Use Your GI Bill Benefits

"ANYONE WHO STOPS LEARNING IS OLD, whether at twenty or eighty. Anyone who keeps learning stays young." -- Henry Ford

Stay young - become a lifelong learner! There are so many reasons to keep learning. And many of you have the advantage of great educational benefits - especially the Post-9/11 GI Bill. The right education can open doors of opportunity for your civilian career. Advanced education or additional training can make you more competitive in the job market, and can also pay off down the road in earnings.

Consider these questions:

- *Do you need to update your skills or obtain new, more marketable ones?*
- *Are there "soft skills" that would be beneficial to have, in addition to the "hard skills" that are necessary to do a certain job?*
- *Is a license required to work in your desired field?*
- *Would a college degree provide more opportunities for you in coming years?*
- *Would Six Sigma certification improve your chances of working in healthcare, financial services, or information technology?*
- *Would you be more marketable if you added a Project Management Professional credential to your years of project management experience?*
- *Do you want to improve your communication skills or become more comfortable at public speaking?*

Education is an investment of time and money that will pay dividends for the rest of your working life. According to the U.S. Bureau of Labor Statistics, full-time workers age 25 and over with a bachelor's degree can expect to earn 1.7 times more than their counterpart with a high school diploma ($1,248 per week vs. $746 per week). College educated people are also significantly less likely to be unemployed (2.2% vs. 4.1%).

Unemployment rates and earnings by educational attainment, 2019

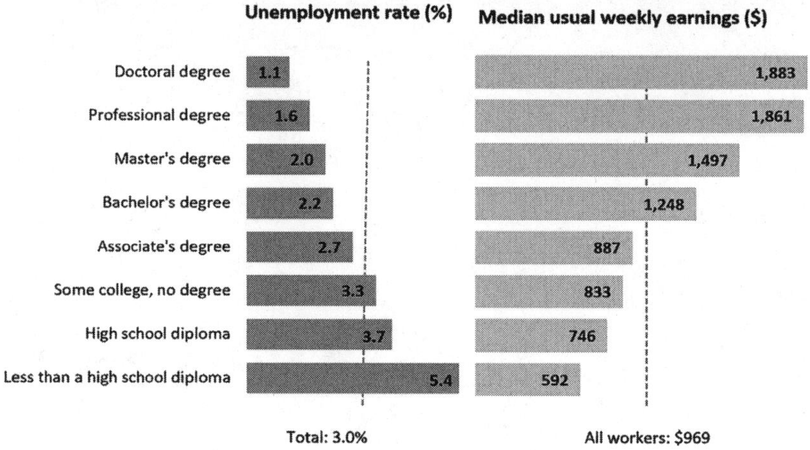

Note: Data are for persons age 25 and over. Earnings are for full-time wage and salary workers.
Source: U.S. Bureau of Labor Statistics, Current Population Survey.

However, you don't have to attain a degree or the highest level of education in order to get a return on your investment. The chart above from DoL's Bureau of Labor Statistics website (https://www.bls.gov/emp/chart-unemployment-earnings-education.htm) does not take into account the completion of training programs, certifications, apprenticeships, and other on-the-job training, which may also positively influence earnings and employment.

Don't hesitate to think outside the box. Look at all possibilities – public universities, private colleges, training and certification programs. A workshop or a seminar can be very effective without being time-consuming. Check out associations in your field that have meetings, conferences, and other events. Join organizations like Toastmasters and subscribe to industry magazines and newspapers. There are many virtual (online) options now, making this even more accessible.

Choose Wisely

Be sure to pursue training that aligns well with your industry and will give you a leg up when employers are comparing you to other candidates. It's important to identify what certifications that civilians in your industry hold, so you know what you need to compete for those civilian positions. To assist you in identifying civilian credentialing requirements for your occupation, visit Credentialing Opportunities On-Line (COOL) as mentioned in Chapter 4. Also check out the credentialing information on DoL's CareerOneStop website.

If you are considering a career change, choose an occupation that is in demand. The Department of Labor's projected employment statistics to the year 2029 identify the fastest growing occupations. Their projections confirm that the best jobs in the future will all require greater investments in education and training. For more information on job trends, see the Bureau of Labor Statistics' Occupational Outlook Handbook.

Do Your Research

Depending on your circumstances, evaluate whether a local or online program, as well as a part-time or full-time basis is best suited to your needs. There are many online degree programs that allow you to pursue your education remotely. Also, many schools offer credit for your military experience.

Do your research on schools and training programs. Choose the right school for your particular discipline. Use the National Center for Education Statistics' College Navigator to search schools by location, programs, and more.

The Department of Veterans Affairs and The Federal Trade Commission both recommend that you consider the following questions when choosing a school:

- Would the employers in my professional field respect a degree from the university or college I'm considering?
- How well does the school support veterans?
- Will I get credit for my military training?
- What is the school's graduation rate and retention rate?
- What's the total cost?
- How will I pay for tuition and expenses that my VA benefits don't cover?
- Will a degree from this school get me where I want to go?
- Is the school accredited?
- What percentage of graduates is late in paying back loans?
- What is the average starting salary of its graduates?

In addition to traditional institutions of higher education - colleges and universities that grant undergraduate and graduate degrees - the Post-9/11 GI Bill provides assistance for many other types of education/training:

- Accelerated payments for high-tech programs
- Correspondence training
- Cooperative training
- Entrepreneurship training
- Flight training
- Independent and distance learning
- Licensing and certification reimbursement
- National testing reimbursement
- On-the-job training and apprenticeships
- Tuition Assistance top-up
- Tutorial assistance
- Vocational/Technical training, non-college degree programs
- Veteran Employment Through Technology Education Courses (VET TEC Program)

You can search for approved education and training programs, eligibility criteria, and license and certifications programs on the GI Bill website (www.benefits.va.gov/gibill).

The Time is Now!

There is currently a wide-spread effort by schools to accommodate active-duty service members and veterans. More schools are accepting American Council on Education (ACE) credits, which convert military training into academic hours. Schools are expected to increase the number of online courses and degree programs, which gives you more flexibility. And thanks to the Veterans Access, Choice, and Accountability Act of 2014, all VA-approved programs in public colleges and universities will soon offer in-state tuition to covered veterans.

If you are considering a private institution where tuition is higher than that of an in-state program, check to see if your school participates in the Yellow Ribbon Program, which allows for schools to contribute additional funds for your education program (with a matching amount from VA directly to the school) without an additional charge to your GI Bill entitlement.

The "Forever GI Bill"

Formally known as the Harry W. Colmery Veterans Educational Assistance Act of 2017, this "Forever GI Bill" changes the Veterans education benefits in some significant ways.

- Elimination of 15-year limitation to use the Post-9/11 GI Bill Program for veterans who left active duty on or after January 1, 2013 and for some Fry Scholarship children and spouses
- Use of GI Bill benefits at an accredited independent study program at career and technical schools and postsecondary vocational schools (previously programs were not approved if any portion was online)
- Restoration of benefits and relief to veterans affected by school closures and disapprovals
- Permanent authorization of qualifying work-study activities
- Identification of schools offering priority enrollment to veterans
- Consolidation of benefit levels: eliminates the 40-percent benefit level and expands the 60-percent level
- More benefits for Science, Technology, Engineering and Math (STEM) Programs: up to an additional 9 months of benefits to eligible individuals
- Pilot Programs for Technology Courses (VET TEC)
- Changes to Licensing and Certification Charges: prorated based on the actual amount of the test fee
- Purple Heart Recipients: Servicemembers and honorably discharged Veterans who were awarded a Purple Heart on or after September 11, 2001 will be entitled to Post-9/11 GI Bill benefits at the 100-percent benefit level for up to 36 months
- Yellow Ribbon Extension to Fry Scholarship and Purple Heart Recipients
- Monthly Housing Based on Campus Student Attends the Majority of their Classes,
- Reserve Duty That Counts Toward Post-9/11 Eligibility
- Changes to Survivors' and Dependents' Educational Assistance
- Informing Schools about Beneficiary Entitlement
- Reserve Component Benefits and Monthly Housing Allowance
- GI Bill Monthly Housing Allowance
- Assistance for Post-9/11 GI Bill students impacted by Colmery Act MHA changes
- Yellow Ribbon Extension to Active Duty Servicemembers (to be effective 8/1/2022)

Put your future on track for success and boost your job and earning opportunities by using your military education benefits today! And never stop learning.

For links to the resources listed in this chapter,
visit www.CorporateGray.com > Transition Guide > Use Your Education Benefits.

chapter 6

Create Dynamite Resumes and Letters

NOW THAT YOU KNOW (1) what you do well, (2) what you enjoy doing, and (3) what you want to do in the future, you have the basic information necessary for communicating your qualifications to employers. But what will you do with this information? What messages do you want to send to employers about yourself? How do you plan to convey these messages—by phone, letter, e-mail, or in face-to-face in-person or virtual meetings?

Will a prospective employer be motivated to learn more about you from reading your resume? Resumes are important tools for communicating your purpose and capabilities to employers. A resume advertises your qualifications and is your calling card for getting interviews. You must craft your resume by thinking through your objective and understanding the different types, construction elements, and refinements of a resume that are necessary to make it an effective first impression to employers.

Define Your Goals and Objectives

When writing a resume, you can choose to include an objective either on your resume or in the body of your cover letter. Including an objective on your resume is optional, but having an objective is not, as it forces you to focus on your employment goal. Objectives are statements of what you want to do in the future. When combined with a skills and value assessment, objectives give your job search needed direction and meaning.

When you identify your strengths, you create the necessary database and vocabulary for developing your job objective. Using this vocabulary, you will be able to effectively communicate to employers that you are a talented individual who achieves results.

Including a job or career objective in your resume that is relevant to your skills and your employers needs tells an employer what it is you want to do, can do, and will do for them. The position you seek is "what you want to do;" your qualifications are "what you have to offer." Your objective should state your strongest qualifications for meeting the employer's needs and should be employer-centered, incorporating both a skill and an outcome in reference to your major strengths and the employer's needs. This objective follows a basic job-skill-benefit format:

I want a [position/job] where I will use my [skills/abilities], which will result in [outcomes/benefits].

For example:

A position in data analytics where skills in mathematics, computer programming, and deductive reasoning will contribute to new systems development.

The following are examples of weak and strong objective statements:

Weak Objectives

- *A position in social services which will allow me to work with people.*
- *A position in Personnel Administration with a progressive firm.*
- *Sales Representative with opportunity for advancement.*

Strong Objectives

- *A public relations position which will maximize opportunities to develop and implement programs, to organize people and events, and to communicate positive ideas and images.*
- *A position as a General Sales Representative with a pharmaceutical company which will use my chemistry background and ability to work on a self-directed basis in managing a marketing territory.*
- *Responsible position in investment research and analysis. Interests and skills include securities analysis, financial planning, and portfolio management. Long range goal: to become a Certified Financial Planner.*

It is important to relate your objective to your audience. Your audience wants to know what you can do for them, so keep your objective work-centered rather than self-centered.

Use Resume Assistance Designed for You

We realize this may be your first resume upon leaving the service. Because it is so important to your job search, your resume needs to be well crafted with both your goals and the employers' needs in mind.

We recommend that you initially draft the resume on your own. By drafting the first version yourself, you will be forced to organize your thoughts in regard to the jobs you've held and your accomplishments in each. In the final analysis, the resume represents you—and who knows you better than yourself?

After you've drafted your resume, meet with the career counselors at your local transition assistance office. These people have reviewed hundreds of resumes and will be able to expertly evaluate yours. Remember, you are looking for honest feedback rather than a pep talk. These specialists are professionally trained in the art of resume writing. They can help ensure your resume effectively communicates your qualifications.

You can also obtain assistance in revising your resume through some of the federally sponsored programs identified in Chapter 2. There are many military service, federal government, and association-sponsored transition services available to assist you with your resume at no or minimal expense. Be sure to take advantage of these numerous resources. Many of these resources are also available to military spouses.

Communicating Positive Images

At every stage in the job search you must communicate a positive image to potential employers. The initial impression you make on an employer will determine whether the employer is interested in interviewing you and ultimately offering you a position.

Unfortunately, many resumes exhibit these characteristics, which are often reported by employers as major weaknesses:

- Unfamiliar military acronyms
- Poor layout
- Misspellings and punctuation errors
- Poor grammar
- Unclear purpose
- Too much military jargon
- Irrelevant data
- Too long or too short
- Poorly typed and reproduced
- Unexplained time gaps
- Too boastful
- Lacks permanent contact information
- Difficult to understand or interpret
- Overstating one's qualifications

Keep in mind that most employers are busy people who normally glance at a resume for only 20 or 30 seconds. Your resume, therefore, must sufficiently catch their attention. When writing your resume, ask yourself the same question a hiring manager would ask: "*Why should I read this resume or contact this person for an interview?*" Your answer should result in a clear, concise, easy-to-read, and skills-based resume.

Types of Resumes

There are three basic types of resumes to choose from: chronological, functional, and combination. Each format has various advantages and disadvantages, depending on your background and purpose. The sample resumes at the end of this chapter give you a template for creating your own. These resumes are shown on the MY RESUME page of Corporate Gray Online (www.CorporateGray.com/resumes/guest_index) and are downloadable as Microsoft Word documents.

Chronological Resume

The chronological resume is the standard resume used by most applicants. It lists your jobs in reverse chronological order with a description of what you did in each job. It shows the progression of your skills and experience. You should use this type of resume when you have extensive experience directly related to a position you seek. This resume should include a work objective which reflects both your work experience and professional goals.

The work experience section should include the names and addresses of former employers followed by a brief description of your accomplishments, skills, and responsibilities. Inclusive employment dates should appear at the end. Be sure to stress your accomplishments and skills rather than your formal duties and responsibilities. You want to inform your audience that you are a productive and responsible person who gets things done—a doer. Military personnel with a great deal of progressive work experience relevant to the civilian work world should use this type of resume. However, for this resume to work best, you must ensure your military experience has been clearly translated into civilian terms.

Functional Resume

The functional resume describes your core competencies and the functions you can perform. It focuses on what you can do outside the context of specific jobs you've held. Functional resumes should be used by individuals making a significant career change, first entering the workforce, or reentering the job market after a lengthy absence. This resume format is designed to stress your accomplishments and transferable skills regardless of previous work settings and job titles. Names of employers and dates of employment should not appear on the functional resume.

Functional resumes have certain weaknesses. While they are important bridges for the inexperienced and for those making a significant career change, many employers dislike this resume format as they want to see what, where, and when you did the work. Use a functional resume only if you have limited work experience or your past work experience does not strengthen your objective when making a major career change.

Combination Resume

The combination resume is a hybrid that combines your functional expertise with your work history. Having more advantages than disadvantages, this resume may be exactly what you need as you transition to a civilian career.

Combination resumes have the potential to both meet and raise the expectations of employers. It enables you to stress your accomplishments and skills, as well as include a synopsis of your work history. This is the appropriate resume format for someone with many years of military experience who wants to accentuate and group those work elements that best map to their desired civilian occupation.

For most retiring military, we recommend the combination format because it highlights your functional expertise and also shows your most recent assignments in reverse chronological order. However, many companies, especially defense contractors, want to see a chronological resume because it provides an unbroken chain of work experience coupled with your levels of responsibility and accomplishments. Examples of the different types of resumes are included later in this chapter.

Constructing Your Resume

In Chapter 4 you documented your experience and accomplishments for each assignment. Use these worksheets to communicate your experience, work history, and education to employers. Read through what you recorded earlier and consider how your experience and accomplishments could be stated in terms that civilian readers will understand.

What you should or should not include in your resume depends on your particular goals, as well as your situation and the needs of the employer. At the very least, your resume should include the following five categories of information:

Contact Information – Who you are/how to contact you: your name, city, state, phone number, email address, and LinkedIn profile URL.

Objective – A career objective relevant to your skills and the employer's needs.

Summary of Qualifications – What you can do – your patterns of skills and accomplishments.

Work History – What you have done – your job titles and activities performed.

Education – What you have learned – your education and training/degrees and certifications.

Other categories may be included but should be limited to only those relevant to your job search, such as professional affiliations, special skills, awards, and special recognition.

Use the information you've collected to create a template in chronological resume format. Start with the **Contact Information**—your name, address, phone, and e-mail.

Second, under **Objective**, enter your employer-centered job or career objective, if you've chosen to include this element in your resume.

Third, under **Qualifications Summary**, develop a short (5- or 6-line) paragraph or put down a few bulleted items that highlight your best selling points.

Fourth, starting with your most recent assignment and working backwards, enter the different assignments (**Employment History** or **Professional Experience**) you held, along with a brief description of your responsibilities and, more importantly, your accomplishments. Where possible, try to quantify an accomplishment in a way that shows you did something measurably better, faster, or cheaper.

Fifth, include relevant **Education and Training**. Finally, if you have **Personal Data** relevant to your objective, put it down. For example, if you want to be a Russian translator and you speak Russian fluently, be sure you include this fact under Personal Data. Be careful to include only information designed to strengthen your objective.

You have just created the first draft of your resume. That wasn't too difficult, was it? While your first draft may run more than two pages, try to get everything into one or two pages, as most employers lose interest after reading the first two pages.

Your final draft should conform to the following do's and don'ts:

Resume Do's

- Do include an employer-centered objective.
- Do focus on those accomplishments that relate to the employers' needs.
- Do use action verbs and the active voice.
- Do include nouns so your resume can be scanned for keywords.
- Do be direct, succinct, and expressive with your language.
- Do appear neat, well organized, and professional.
- Do use ample spacing and highlights (all caps, underlining, bulleting) for different emphases (except if it's an electronic resume).
- Do maintain an eye-pleasing balance. Try to center your contact information at the top, keeping information categories on the left in all caps.
- Do check carefully your spelling, grammar, and punctuation.
- Do clearly communicate your purpose and value to employers.
- Do communicate your strongest points first.
- Do seek reviews.
- Do refine, rewrite, and proofread several times.

Resume Don'ts

- Don't use military acronyms or slang.
- Don't use abbreviations except for your middle name.
- Don't make the resume cramped and crowded.
- Don't make statements you can't document.
- Don't use the passive voice.
- Don't change the tense of verbs.
- Don't use lengthy sentences and descriptions.
- Don't refer to yourself as "I."
- Don't include negative information.
- Don't include salary information (unless required).
- Don't include a photograph of yourself.
- Don't include extraneous information.

Refining the Final Product

Resume Sequence

What is most important to both you and the employer? Your most important information and strongest qualifications should come first. If your education is more relevant to your next job than your work experience, then the sequence of elements should be:

- Contact Information
- Objective
- Education
- Summary of Qualifications
- Work History

If you have several years of direct work experience that support your objective, and if education is not an important qualifying criteria, then the sequence should be:

- Contact Information
- Objective
- Summary of Qualifications
- Work History
- Education

Resume Critiquing

Does your resume accurately present your qualifications in the best possible light so that a hiring manager will want to call you in for an interview? That is the true test of an effective resume!

You should subject your resume drafts to two types of evaluations. An **internal evaluation** consists of reviewing our lists of "do's and don'ts" to make sure your resume conforms to these rules. An **external evaluation** should be conducted by circulating your resume to three or more individuals whom you believe will give you objective and useful feedback (avoid people who tend to flatter you). Ask that they critique your draft resume and suggest improvements in form and content. The best evaluator would be someone in a hiring position similar to one you will encounter in the actual interview. Asking someone to critique your resume is one way to spread the word that you are job hunting. As we will see in Chapter 8, this is one method for getting invited to an interview!

Paper Versus Electronic Resumes

Employers generally receive resumes in one of three ways—as a paper copy, as an electronic (digital) resume (e.g., Microsoft Word, Adobe PDF), or by downloading the candidate's resume from a database. The resume format that you use will depend on the circumstance. For example, if you are going to attend an in-person job fair, you will want to bring paper copies. If, on the other hand, you are searching the Internet and find an opportunity of interest, you will want to send an electronic resume, most likely in Microsoft Word or Adobe PDF format.

Another form of the electronic resume is submitting it to a company's website. Generally this will require you to either upload your resume as a file or "copy and paste" your resume into a standard electronic form and then enter additional pieces of data such as security clearance level, education level, willingness to relocate, etc.

In addition to the company website, there are resume databases organized specifically to facilitate the hiring of military personnel by civilian organizations. Employers search these resume databases to identify potential job candidates. One such military-specific site is Corporate Gray Online (www.CorporateGray.com). There are also general employment websites such as Monster (www.Monster.com), CareerBuilder (www.CareerBuilder.com), and Indeed (www.Indeed.com) where you can upload your resume for recruiters to search.

Preparing Your Electronic Resume

Distributing your resume via the Internet has many advantages: it is fast, it reaches employers world-wide, and it is free. Here are some steps you can take to ensure your resume converts neatly to html for clean electronic uploading and sharing:

- **Use Microsoft Word** - Save it in the .doc or .docx format of Microsoft Word. You can also save your resume in PDF format for uploading.
- **Keep it simple** - This means no headers, footers, graphics, or tables. A good check for clarity is to convert your document to plain text and see how it reads ("Save as" plain text *.txt).
- **Capitalize headings** - Putting the headings in all capital letters will help differentiate the sections of your resume. Save caps for just the main headings (EXPERIENCE, EDUCATION, etc.) and use them sparingly so as not to clutter the resume's readability.
- **Use blank lines sparingly** - When a document is uploaded, blank lines are often interpreted as separation between sections. Use blank lines to separate sections and differentiate job entries, but don't include blank lines within a single job description as it could appear disjointed when uploaded.

Companies will frequently scan and OCR the paper resumes they receive and store the resume in their database for later retrieval. Since a scanned file can be problematic, you will want to send them an electronic version of your resume as well.

Final Production

We recommend creating your resume using Microsoft Word and printing it using a high quality laser printer. When printing the resume, consider the quality and color of paper. We recommend watermarked 20-pound or heavier bond paper in either off-white or light cream. The text color should be black. If your resume is more than one page, put your name at the top of each page. Do not use staples or report covers to bind the sheets.

Whatever choices you make, do not cut costs when it comes to producing your resume. Remember, your resume is your corporate calling card—it should represent you in a professional manner. Put your best foot forward by producing a first-class resume.

Job Search Letters

Letters also play a key role in a job search. They come in different forms, from cover letters to approach and thank-you letters. Mailed or transmitted resumes are normally accompanied by a cover letter. After interviewing for information or a position, you should send a thank-you letter, another type of job search correspondence. Other occasions will arise when it is both proper and necessary for you to write different types of job search letters. There are examples of these letters at the end of this chapter. These job search letters are shown on the MY RESUME page of Corporate Gray Online (www.CorporateGray.com/resumes/guest_index) and are downloadable as Microsoft Word documents.

Job hunting letters are like resumes—they advertise you for interviews. Like good advertisements, these letters should follow four basic principles for effectiveness:

1. Catch the reader's attention.
2. Persuade the reader of your benefit or value.
3. Convince the reader with factual evidence.
4. Move the reader to acquire the product.

Basic Preparation Rules

Before you begin writing a job search letter, ask yourself several questions to clarify the content of your letter:

- What is the purpose of the letter?
- What are the needs of my audience?
- What benefits will my audience gain from me?
- What is a good opening sentence for grabbing the attention of my audience?
- How can I maintain the interest of my audience?
- How can I best end the letter so that the audience will be persuaded to contact me?
- If a resume is enclosed, how can my letter best advertise the resume?

Since your letters are a form of business communication, they should conform to the rules of good business correspondence:

- Plan and organize what you will say by outlining the content of your letter.
- Know your objective, and structure your letter accordingly.
- Communicate your message in a logical and sequential manner.
- State your purpose immediately in the first sentence and paragraph;
- Use short paragraphs and sentences; avoid complex sentences.
- Punctuate properly and use correct grammar and spelling.
- Use simple and straightforward language; avoid jargon.
- Communicate your message as directly and briefly as possible.
- Indicate what follow-up actions you will take.
- End by stating what your reader can expect next from you.

The rules stress how to both organize and communicate your message with impact. At the same time, you should always have a specific purpose in mind as well as know the needs of your audience.

Types of Letters

Cover letters provide cover for your resume. You should avoid overwhelming a one-page resume with a two-page letter or repeating the contents of the resume in the letter. A short and succinct one-page letter which highlights one or two points in your resume is enough.

The first paragraph should state your interest and reason for writing. The second paragraph should highlight your possible value to the employer. The third paragraph should state that you will call the individual at a particular time to see if an interview can be scheduled.

Approach letters are written to develop job contacts, leads, or information as well as for organizing networks and getting interviews—the subjects of Chapter 3. These letters help you gain access to the hidden job market by making networking contacts that lead to those all-important informational interviews. Your primary purpose should be to get employers to engage in the 5R's of informational interviewing, which is discussed in more detail in Chapter 3:

- Reveal useful information and advice.
- Refer you to others.
- Read your resume.
- Revise your resume.
- Remember you for future consideration.

Approach letters can be sent out en masse to uncover job leads, or they can target particular individuals or organizations. It is best to target these letters since they have maximum impact when personalized in reference to particular positions.

The structure of approach letters is similar to other letters. The first paragraph states your purpose. In so doing, you may want to use a personal statement for openers, such as "*Mary Tillis recommended that I write to you...*" or "*I am familiar with your...*" State your purpose, but do not suggest that you are asking for a job—only career advice or information. In your final paragraph, request a meeting and indicate you will call to schedule such a meeting at a mutually convenient time.

Thank-you letters may become your most effective job search letters. They communicate your thoughtfulness. These letters come in different forms and are written for various occasions. The most common thank-you letter is written after receiving assistance, such as job search information or a critique of your resume. Other occasions include:

- **Immediately following an interview:** Thank the interviewer for the opportunity to interview for the position. Repeat your interest in the position.
- **After receiving a job offer:** Thank the employer for his or her faith in you and express your appreciation.
- **After being rejected for a job:** Thank the employer for giving you the opportunity to interview for the job. Ask to be remembered for future openings.
- **When terminating employment:** Thank the employer for the experience you gained while working at their company.
- **Beginning a new job:** Thank the employer for giving you this new opportunity and express your confidence in producing the value he or she is expecting from you.

Though these thank-you letters are for different occasions, they all have the same goal in mind—to be remembered by potential employers in a positive light. A thank-you letter is a powerful way to get remembered.

Distribution and Management

Good resumes are those that get read, remembered, referred, and result in a job interview. Therefore, after completing a first-rate resume, you must decide what to do with it. Are you planning to only respond to classified ads with a standard mailing piece consisting of your paper or electronic resume and a formal cover letter? Do you prefer posting your resume online with resume databases or e-mailing it to potential employers? But wait a minute; classified ads and electronic databases only represent one portion of the job market. What other creative distribution methods might you use, such as sending it to friends, relatives, and former employers? What is the best way to proceed?

Responding to Classified Ads

While electronic media is the most prevalent, you will still see classified ads being used to announce job openings. There are a few key principals to keep in mind when responding to classified ads:

- Be selective and only respond to those ads that are relevant to your job search.
- Respond with a cover letter and resume as soon as possible. Your letter should be

tailored to emphasize your qualifications per the position requirements, using similar terminology to emphasize your qualifications.
- Use your judgment in addressing the salary question. There is no hard and fast rule on stating a figure or range, but it's usually best to say negotiable. You should defer salary until the end of the interview—after you have demonstrated your value and have more information about the position. This is discussed this further in Chapter 9.
- Since employers are often swamped with responses initially, you may want to send another copy of your resume and cover letter two or three weeks later.
- Don't spend an excessive amount of time responding to one ad. Your goal should be to contact as many relevant employers as possible, because uncovering fruitful job leads is a numbers game.

The Internet

One of the easiest, most cost-effective ways to distribute your resume is online. It has some obvious advantages. First, it's fast. You can get your resume in the hands of employers in a matter of seconds rather than days through the mail. Second, by providing this information electronically, you are enabling prospective employers worldwide to review your qualifications and match them against your needs. Third, it's free!

If you are targeting specific companies, be sure to check out their websites for employment information and tips. Many company websites include an employment section.

Military-Specific Employment Websites

There are resume databases organized specifically to facilitate the transition of military personnel into the civilian workforce. Prospective employers search these resume databases to quickly identify potential job candidates.

Several companies operate military-focused online employment sites. For example, the Corporate Gray website (www.CorporateGray.com) lets you upload your resume for employers to view. It also has thousands of employment opportunities posted by military-friendly companies and government agencies. We also encourage you to visit other military-specific sites where you can post your resume and acquire job search information.

Civilian Employment Sites

There are also many civilian employment sites designed to link employers with job seekers where you can post your resume and browse thousands of job postings. Many of these sites also include useful job search information and advice.

So which sites will be most useful in your job search? We recommend starting here:

- Indeed www.indeed.com
- Monster.com www.monster.com
- CareerBuilder www.careerbuilder.com
- CareerOneStop - Job Bank www.jobbankinfo.org
- Jobs.com http://jobs.com

Social Media Sites

You can also share your resume on your social media sites, such as LinkedIn (www.linkedin.com). Consider attaching your resume to the Featured or Experience sections of your LinkedIn profile. If you want to attach more than one version of your resume, title each document so it is recognizable by recruiters viewing your profile. Let them know from the title that it is worth downloading. For example, Smith-Contract-Specialist.pdf.

On Facebook (www.facebook.com) you can use the Jobs functionality or add applications such as BranchOut or BeKnown that are designed to build your professional profile and help you search for jobs.

Learn more about how to use social media in your job search from the Corporate Gray Blog (www.blog.corporategray.com) or the book, *Social Media and Your Job Search: Maximizing Your Network for a Successful Transition*, available for purchase on the Corporate Gray Online Transition Guide page at www.CorporateGray.com/static/transition_guide/overview.

Record Keeping

Once you begin distributing letters and resumes, you will need to keep good records for managing your job search campaign. Purchase file folders for your correspondence and be sure to keep copies of all letters since you may need to refer to them when speaking to employers over the phone.

Record your activities with each employer—letters, resumes, phone calls, interviews—either electronically with a spreadsheet (such as Microsoft Excel), using the Company Contacts worksheet introduced in Chapter 7, or setting up a card system organized according to the name of the organization or individual. These files will help you quickly access information and evaluate your job search progress.

Always remember the purpose of resumes and letters—*to advertise you for interviews.* Since most employers know nothing about you, you must effectively communicate your value in writing prior to the critical interview.

Resume and Letter Examples

The following resumes and letters incorporate the writing principles outlined in this chapter. These sample resumes and letters are downloadable as Microsoft Word files on www.CorporateGray.com/resumes/guest_index. Additional resume and cover letter examples are found in the guide *Military-to-Civilian Resumes and Letters* (Impact Publications).

Remember, your resume and letters are your personal calling cards. They should convey your qualifications in a manner that catches the reader's attention and quickly illustrates how your skills and experience can be used to a prospective employer's advantage.

For a handy checklist of Resume Do's and Don'ts and to download sample resumes and letters, visit www.CorporateGray.com > Transition Guide > Construct Your Resume.

Resume Anatomy 101

These are the elements of an effective resume. Include keywords to describe your skills and experience, and tailor your resume for each job you are targeting.

Contact Info
Include your name, city, state, phone, email, and LinkedIn profile URL

Objective
It is optional to include your mission statement – make it concise

Qualifications
List your top skills and qualifications – tailor the list to the job you are targeting

Experience
Translate your skills into civilian speak – use keywords recruiters will search

Education
This could be placed above experience, depending on its importance to the job you are targeting

Awards & Honors
Include honors you've received that are relevant to your objective

Links
Optional – add links to your online portfolio, articles you've written, etc.

DANA PHILLIPS
Madison, CT | 233-245-6789
dphillips@gmail.com | linkedin.com/danaphillips

OBJECTIVE
Human Resources position where military recruiting and leadership experience can be used to enhance the operational effectiveness of a company's HR operations.

QUALIFICATIONS
Skilled communicator with proven personnel administration expertise. Knowledgeable in all facets of HR management.

- Staffing
- Recruiting & Retention
- Benefits
- Training
- Legal Issues
- Possess a Secret Clearance

EXPERIENCE
Human Resource Manager: *Pentagon, Washington, DC, 20__ - Present*
Planned, organized, and managed daily activities of a Human Resources department managing the staffing requirements and availability of 500 sea-based enlisted Navy billets.
Raised the efficiency of the department's personnel management operations by 25%.

Branch Chief: *Enlisted Military Affairs, Washington, DC, 20__-20__*
Responsible for daily personnel administration activities, including the creation of orders for Navy personnel moving to new duty locations, awards, promotions orders, disciplinary paperwork. Ensured all paperwork was completed accurately and on time.

EDUCATION
Associate of Arts in Marine Science, *University of Maryland, 20__*

HONORS
Received Commanders Award for outstanding service, 20__
Navy Achievement Medal, 20__

LINKS
www.danaphillips.com
www.hrreview.com/article-by-phillips

Junior Officer

Sanford J. Foxx

Ft. Dix, NJ
(609) 321-4545

SJFoxx@gmail.com
linkedin.com/sjfoxx

EXPERIENCE

Captain, Field Artillery: Graduate U.S. Army Airborne School

May 20__-present: Company Commander **Ft. Ord, California**
Led a diverse group of 130 soldiers to prepare for military missions. Worked through senior enlisted personnel to ensure training assignments and operations were completed according to schedule. Established the organizational climate that fostered teamwork and respect for others. Evaluated the performance and set the work standards for unit members. Assigned work responsibilities based on ability of individuals. Allocated human and financial resources to ensure accomplishment of unit mission.

- Under my leadership, unit was recognized as "Outstanding" for three consecutive years during the annual training evaluation cycles.
- Improved unit operational effectiveness by 30% and increased unit retention by 15%.

20__-20__: Deputy Logistical Staff Officer **Joint Base Mc Guire Dix, NJ**
Staff officer responsible for national logistical support operations and selected international locations.

- Built coalitions with civilian and governmental agencies which increase support by 34%.
- Worked with diverse group of officers to develop new and efficient ways to deliver logistical support.

20__-20__: Deputy Chief of Staff **Fort Sill, OK**
Worked through other Chiefs of Staff to effectively coordinate domestic missions involving 1,500 soldiers. Assigned work to subordinate units and ensured projects were completed on time and on budget. Oversaw the development of the Five Year Strategic War Plans for unit. This plan set the direction for domestic and international warfighting.

- Developed, implemented, and marketed innovative communication plans to speed the distribution of key unit information. These plans increased interunit communication flow by 35%.
- Awarded "Superior Rating" during annual inspection by higher headquarters two years consecutively.

20__-20__: Fire Direction Officer **Fort Lewis, WA**
Led a diverse group of 10 enlisted soldiers to accomplish technical tasks needed for indirect artillery fire. Managed unit capital budget to ensure proper allocation of fiscal resources. Assessed individual performance based on uniform standards. Received 3 citations from higher command for "Outstanding Leadership" and "Highest Unit Morale."

- Recorded highest inspection score for unit readiness for 2 consecutive years.
- Improved unit training effectiveness by 35%.

EDUCATION

U.S. Military Academy, West Point, NY, 20__
- Bachelor of Science; Major in Electrical Engineering
- Summer Intern in Africa
- Four year starter on Men's Varsity Soccer Team

ADDITIONAL:

- Top Secret Security Clearance
- Lean Six Sigma, Black Belt Certified Trained
- Fluent in MS Office Products

Resume provided by Dr. Jerusalem T. Howard, JTH Coaching; http://jthcoaching.com

Logistics

JOHN JOBSEEKER

Fair City, Virginia • 804-555-1111 • johnjobseeker@gmail.com • linkedin.com/johnjobseeker

SENIOR LOGISTICS MANAGER

Dedicated logistics manager with active Secret Security clearance and extensive experience leading diverse teams under a range of adverse conditions. Highly trained in all aspects of supply chain management, including purchasing, budgeting, storage, distribution, and integrated property accounting systems. Positive communicator with outstanding supervisory, training, and coaching skills. Areas of expertise:

Project Management • Supply Chain Management • Procurement • Budgeting • Marketing Skills DoD/Military Standards • Automated Supply Accounting Systems • Contract Administration Process Improvement • Risk Management • Cost Control • Change Management • P&L Professional Development • Training Programs • Coaching • Mentoring

PROFESSIONAL EXPERIENCE

UNITED STATES ARMY, Fort Bragg, North Carolina
Executive Officer • 20__ – Present
Managed and coordinated 120 administrative, intelligence operations, and logistics staff members to perform as a cohesive unit, reporting directly to unit Commander. Administered $1.5M unit budget and ensured maintenance of and accountability for $22M of equipment at all times. Orchestrated training and development of subordinate officers to ensure continuity of leadership.

- Maintained extraordinary equipment operational readiness rate exceeding 95%.
- Ensured budgetary accountability by implementing monthly budget subdivided into weekly amounts, and enhanced financial planning for anticipated future projects.

Senior Logistics Manager • 20__ – 20__
Issued vehicles, night vision equipment, weapons, radios, and other equipment to units deploying to Iraq and Afghanistan. Supported overall operations by anticipating logistical requirements and ensuring the timely completion of both expressed and implied tasks. Ensured all equipment was properly maintained and accounted for using multiple computerized logistics and maintenance programs.

- Performed monthly inventories of general and specialized high-tech equipment, and liaised with internal and external stakeholders to maintain 100% operational readiness.
- Achieved 100% loss prevention while training more than 15,000 soldiers for overseas deployment.
- Prepared and administered contracts procuring equipment training for deployment to Afghanistan.

Operations Officer • 20__ – 20__
Developed and administered training programs for 65 Army Reserve and Army National Guard battalions and brigades. Designed validation process for units deploying overseas. Managed unit status reporting requirements to achieve short- and long-term planning goals. Responded to rapidly changing combat environments by designing program modifications enhancing effectiveness of training.

- Spearheaded training of 20,000 reservists and National Guard soldiers, preparing personnel from diverse **backgrounds to work together and react effectively in emergency situations.**

HHC Company Commander • 20__ – 20__
Managed and supervised 162 soldiers, ensuring fair and consistent training and disciplinary standards to ensure unit cohesion and readiness. Administered disciplinary programs, wrote commendations, issued awards, and coordinated training and development programs. Served as liaison between family members and Brigade Commander to ensure open communication on how current and future operations will affect soldiers and their families.

- Earned commendation from Division Commander for designing innovative live-fire convoy training exercise for units deploying overseas.
- Transformed Family Readiness Group and revitalized support programs serving soldiers' families.

EDUCATION

Bachelor of Arts in Political Science
Virginia Commonwealth University • Richmond, Virginia

Resume provided courtesy of GetInterviews.com; email: resumes@getinterviews.com

Logistics

John Smith
Fort Benning, GA
John.Smith@gmail.com Cell: (123) 555-1234
Linkedin.com/johncsmith2

LOGISTICS OPERATIONS SUPERVISOR

U.S. Government Secret Security Clearance
Military-trained professional with a keen attention to detail and an analytical mind, continually finding ways to improve the overall operations of the organization. Track record of identifying redundancies and maximizing resources to streamline operations. A member of the United States Army Transportation Corps, providing flawless movement of personnel and material by various modes of transportation. Bilingual in Spanish. Core competencies in:

- Large-scale Operations Management
- Safety Management
- Data Analyst
- Logistics Support, Planning & Execution
- Personnel Organization & Supervision
- Department of Defense Compliance

PROFESSIONAL EXPERIENCE

Operations Officer, *U.S. Army, Fort Benning, GA* May 20__ to Present
Plan and arrange the deployment efforts of 5 units consisting of 790 Soldiers in support of Iraq and Afghanistan missions. Analyze current contingency programs and propose and implement improved procedures in order to enhance mission capability. Supported 350 Soldiers by performing legal and personnel actions for each. Conduct comprehensive legal procedures such as investigations in order to make well-informed, impartial, and timely decisions regarding repercussions.

- Enhanced training and leadership development of personnel by implementing a training cycle management program.
- Maintained Soldier readiness in marksmanship, physical training, functional training, and ensured drivers had proper certifications to ensure 100% deployment readiness.
- Increased mission aptitude by providing expert strategic technical skills while recognizing and correcting problems through intensive research and training.

Company Commander, *U.S. Army, Joint Base Balad, Iraq, March 20__ to May 20__*
Directed more than 140 personnel consisting of Soldiers, Airmen and contractors in a multifaceted environment within six different areas in support of Operation Iraqi Freedom (OIF). Administered all transportation operations consisting of 200K personnel and approximately 170K tons of cargo on more than 7K aircraft. Maintained an accurate tracking database of all resources and equipment.

- Seamlessly managed the transfer of eight thousand individuals and one thousand pieces of aircraft through proper preparation of the terminal as well as leadership capabilities.
- Enhanced mission success by implementing operational planning procedures that aided in the transfer of 37K pieces of cargo weighing approximately 14K tons throughout Iraq.

Company Executive Officer, *U.S. Army, Fort Bragg, March 20__ to March 20__*
Championed the training, administration, morale and welfare of 400 personnel in preparation for deployment efforts in support of OIF. Maintained resources valued at $3M and ensured more than 150K Soldiers were trained, practiced and prepared for utilization of equipment and tools. Developed and implemented education on sustainment mission training, training operations and peacetime missions. Detachment commander for 20 personnel dealing with legal actions/accountability for equipment such as vehicles, generators, and computers worth more than $1M of property.

- Driving force in preparing multiple units consisting of 700+ personnel to deploy in support of OIF

Movement Control Team Commander, *U.S. Army, Fort Bragg, May 20__ to March 20__*
Provided guidance and recommendations on all areas of mobilization, training requirements, unit maintenance/operations and the overall readiness of the unit. Planned, coordinated and evaluated logistical actions required to support specified missions in support of Operation Enduring Freedom (OEF). Ensured mission readiness of 70 personnel by providing training, instruction and guidance on all aspects of logistics.

- Completed a seamless movement of approximately 1K tons of cargo and two thousand individuals, alleviating the need of more than 200 containers, which saved the Army $20K per day per container, totaling millions of dollars in savings.
- Enhanced the mission capability of more than 600 missions by providing expert logistical support and guidance to five units.
- "He is a proven trainer, leader and warrior and demonstrates these traits flawlessly during OEF."

EDUCATION

Bachelor of Arts Degree, History, Augusta State University, Augusta, GA, 20__

Resume provided courtesy of Blake Mundy; BMundy@MilitaryResumes.com

Law Enforcement Officer

Available: Immediately

703.111.2222 **John T. King** kingj@gmail.com
Alexandria, VA
linkedin.com/johntking

OBJECTIVE: Government law enforcement position where leadership skills and military police experience will benefit a police force seeking dedicated, culturally-sensitive, law enforcement professionals.

CAREER SUMMARY: Possess extensive Physical Security and Law Enforcement background, including over 7 years experience in security, police patrolling, supervision, and management. Assertive, take-charge leader ready to apply skills in a corporate setting.

WORK EXPERIENCE

Fort Belvoir, Operations NCO, 437th Military Police Company ***20__-Present***

Responsible for managing the daily operations of over 200 Military Police soldiers across a myriad of security-related assignments. Key accomplishments:

Responded to short-suspense taskings with ease and professionalism.

- Ensured right mix of security professionals were deployed to handle all assigned missions.
- Ensured emergency response actions were properly coordinated with local police and state agencies in a timely manner.
- Coordinated, scheduled, and led unit weapon training and field exercises.

Fort Belvoir, Platoon Sergeant, 437th Military Police Company ***20__-20__***

Responsible for the health, welfare, morale, and training of the 32 soldiers in my platoon. Key accomplishments:

- Supervised all daily law enforcement operation.
- Established new suspect apprehension procedures which resulted in a 25% decrease in injuries to law enforcement professionals.
- Implemented computer-based training program that enabled the company to take advantage of "best practice" civilian policing programs.

EDUCATION and TRAINING

A.S., Administrative Justice, Northern Virginia Community College, Annandale, VA 20__

Military Police Basic Course, 20__
Drill Sergeant School (Honor Graduate), 20__
Protective Service Course, 20__
Military Police Instructor Training Course, 20__

Helicopter Maintenance Supervisor

Available: May 1, 20__

Aurelio Rodriguez

San Diego, CA
619-231-3223 RodriguezA@gmail.com
linkedin.com/aureliorodriguez

OBJECTIVE Helicopter Maintenance Supervisor for a company that provides the U.S. Navy with aviation equipment and support.

MILITARY EXPERIENCE

Maintenance Instructor, Naval Base San Diego, 20__ – Present
Instructed and graduated 465 students amassing 220 podium hours in 48 classes in Quality Assurance, Work Center Supervisor, and Maintenance Action Form/Subsystem Capability Impact Reporting Organization Level. Course Curriculum Model Manager of a Chief of Naval Operations approved course in Work Center Supervisor.

Achievements:

- Qualified 4 new instructors in Aviation Maintenance Administration Management.
- Successfully piloted 2 new courses – one in organizational maintenance, another in quality assurance.
- Twice nominated for Instructor of the Quarter.

Maintenance Supervisor, Helicopter Anti-Submarine Squadron 2, San Diego, 20__ – 20__
Supervised 5 personnel in performing scheduled and unscheduled maintenance on the SH-60 F/H helicopter. Performed a wide range of maintenance activities affecting the power plants and transmission systems.

Achievements:

- Received the highest rating possible from higher headquarters following detailed inspection of assigned helicopters.
- Received promotion based on superior performance ratings.
- Earlier assignments were helicopter maintenance-related and of increasing scope and responsibility during the years 20__-20__.

SECURITY CLEARANCE
Top Secret with current Special Background Investigation

EDUCATION
B.S. Aeronautical Engineering, San Diego State University, 20__
3.5 GPA; Dean's List every semester

AWARDS & HONORS
Navy Achievement Medal
Navy Commendation Medal

Human Resources Manager

DANA PHILLIPS

Madison, CT
H: (233) 245-6789 / W: (233) 123-4567
dphillips@gmail.com
linkedin.com/danaphillips

OBJECTIVE

Human Resources position where military recruiting and leadership experience can be used to enhance the operational effectiveness of a company's recruitment and retention program.

QUALIFICATIONS

Skilled communicator with proven personnel administration expertise. Unique blend of managerial and hands-on experience acquired over a career with the U.S. Coast Guard. Knowledgeable in all facets of HR management, including staffing, recruiting and retention, benefits, training, and legal issues. Proficient in Spanish. Possess Secret clearance.

AREAS OF EXPERTISE

Personnel Management

- Planned, organized, and managed daily activities of a Human Resources department charged with managing the staffing requirements and availability of sea-based enlisted Coast Guard billets.
- Used computer models to determine the proper accession numbers, advancement opportunities, and year-end strength goals.
- Recognized by peers and associates for raising the efficiency of the department's personnel management operations.
- Received Commander's Award for outstanding service.

Administration

- Responsible for the daily administration of personnel administration activities, such as the creation of orders for Coast Guard personnel moving to new duty locations or transitioning from the service.
- Processed special personnel actions, including awards, promotion orders, disciplinary paperwork, and marriage applications.
- Ensured all paperwork was completed accurately and within time schedules.

Recruiting

- Served as a front-line human resources recruiter for the U.S. Coast Guard.
- Interviewed, assessed, and recruited candidates interested in military service.
- Recognized for consistently exceeding recruitment goals for quantity and quality over a 24-month period.
- Received Meritorious Service Medal from the East Coast Human Resources Manager.

EMPLOYMENT HISTORY

Human Resource Manager: The Pentagon, Washington, DC, 20__-Present
Branch Chief: Enlisted Military Affairs, Washington, DC, 20__-20__
Detailer: Coast Guard Assignment Branch, Washington, DC, 20__-20__
Seaman: U.S. Coast Guard, 20__-20__

EDUCATION

Associate of Arts, Marine Science, University of Maryland, 20__

Maintenance Technician

ELAINE WILLIAMS

Fairfax, VA
(703) 999-9999

ewilliams@gmail.com
linkedin.com/elainewilliams

MAINTENANCE TECHNICIAN

Over 10 years of experience in the maintenance, repair, troubleshooting and operation of highly complex electronics equipment. Technical skills include hydraulic, pneumatic, mechanical, electrical, and electronic systems. Demonstrated ability to train and lead others to perform productively. Comfortable in fast-paced, high-stress environments requiring attention to detail and ability to meet deadlines within changing priorities.

ACCOMPLISHMENTS

- Completed and achieved the highest number of technical qualifications out of 22 people.
- Selected by senior management as Employee of the Quarter twice in the last year for "superior performance, dedication, professionalism, and positive attitude."
- Recognized as a team player who requires minimum supervision, is motivated to the highest performance standards, and committed to excellence.
- Displayed "unequaled troubleshooting skills in maintenance activities," resulting in the flawless execution of 50+ critical maintenance actions for the safe launch of 10,000+ aircraft.

EMPLOYMENT HISTORY

UNITED STATES NAVY 20__ to Present

Aircraft Maintenance Technician

• Advanced through increasingly responsible positions in aviation equipment operation maintenance, quality assurance, and safety.

• Selected by supervisor out of 22 technicians to handle one of the most complex assignments in the entire division. Exceeded supervisor's expectations.

Equipment Maintenance & Repair

• Experienced in the operation and maintenance of multi-million dollar aircraft.

• Performed troubleshooting of electrical, hydraulic, pneumatic, and mechanical systems.

Quality Assurance & Inspection

• Accurately calibrated and installed 70+ precision measurement tools valued at over $250K.

• Supervised all maintenance checks to monitor accuracy and adherence to exact procedures.

Training & Team Leadership

• Trained and oversaw 40-member work center in the operation & maintenance of equipment.

• Advanced the skills of new personnel, contributing to 15,000+ hours of safe aircraft operations.

EDUCATIONAL TRAINING

Successfully completed specialized training programs in Maintenance Equipment & Operations, Aircraft Firefighting, Catapult Hydraulics, Arresting Gear Hydraulics, and Quality Assurance.

Resume provided by Louise Garver; Louise@careerdirectionsllc.com

Combat Engineer

CHRIS P. JACOBS

Annapolis, MD
*(h) 410-555-1234 ** (c) 410-666-1234*
cpjacobs@gmail.com | linkedin.com/chrisjacobs Available: June 1, 20__

SUMMARY OF QUALICATIONS

• Six years Active Duty U.S. Army -- Sergeant E-5 Team Leader.
• Strong leadership qualities; take charge and manage projects to completion.
• Creative and resourceful in generating new ideas and solving complex problems.
• Takes initiative. Motivates others to perform at highest levels. Leads by example.
• Background and experience in construction & demolition; military-certified.
• Secret security clearance

Quote from recent Army Performance Evaluation

Outstanding leader; able to accomplish any mission. Possesses the moral courage to do the right thing at all times. Places unit's mission and welfare of his soldiers above personal needs.

SKILLS & EXPERIENCE

SUPERVISORY / LEADERSHIP

• Immediately promoted upon passing the rigorous Sergeant's Board evaluation process.
• Directly supervised, trained, and mentored 10 soldiers.
• Led team on over 200 combat patrols earning the respect and loyalty of fellow soldiers.
• Maintained direct accountability for more than $300,000 in military supplies and equipment.

COMMUNICATION / TRAINING

• Conducted 15 classes for the 3rd Brigade Combat Team on military weapons training.
• Provided concise briefings to unit regarding mission, tasks, and objectives.
• Cross-trained personnel as a means of enhancing security and emergency readiness.

WORK HISTORY

Fort Carson, CO	Combat Engineer, Team Leader	June 20__- Present
Fort Leonard Wood, MO	Combat Engineer, Member of Squad	July 20__-2014

EDUCATION & TRAINING

Denver Community College, Denver, CO -- 25 Credit Hours, 20__, Construction Management Diploma
Kendall High School, Jacksonville, FL, 20__, High School Diploma

AWARDS & HONORS

Army Commendation Medal, Presidential Unit Citation, National Defense Service Medal

Resume provided by Beth Colley; resume@chesres.com

Network Administrator

JARROD NOBLE

Fayetteville, NC // (910) 999-9999 // JarrodNoble@gmail.com // linkedin.com/jarrodnoble

Award winning and highly accomplished Network Administrator with proven track record of reducing operating expenses and increasing productivity.

HIGHLIGHT OF QUALIFICATIONS

- Nine years of experience in network administration & technical troubleshooting.
- Outstanding record of resourceful cost reduction.
- Reputation for exceptional leadership and for consistently performing at the highest level.
- Knowledgeable in wide variety of networking environments and software applications.
- Team player with excellent interpersonal and communication skills.

COMPUTER SKILLS

Operating Systems: Windows NT 4.0 and 3.51, Windows for Workgroups 3.11, WANG COBOL

Software Applications: MS Office 2000, MS Exchange and MS Outlook, MS Internet Explorer

Networking Protocols: TCP/IP, NetBui

Hardware: Personal computers and laptops, backup domain controllers, print servers, HP Laser-jet printers, high speed printers, and related peripherals: DVD/CD ROMs, scanners, modems, etc.

NETWORK ADMINISTRATION

Responsible for all software and hardware utilized by squadron personnel. Key accomplishments include:

- Successfully maintained, troubleshot, and repaired array of hardware & software applications.
- Oversaw all user accounts, assigned security rights, and provided desktop support.
- Prepared and coordinated of all data file transfers with local and outside agencies.
- Fully accountable for over 400 pieces of computer equipment valued at over $1.5 million.
- Provided PC configuration, procurement and installation of over 100 desktop computers.
- Dramatically improved computer peripheral inventory through aggressive inspection schedule.

MANAGEMENT & TRAINING

Responsible for all training of computer systems, software applications, local and wide area networks, and peripheral equipment. Key accomplishments include:

- Currently manage mainframe and Windows NT computer systems supporting 150 users.
- Supervised and trained over 50 individuals in Mainframe Information Management Systems.
- Saved thousands of dollars by establishing training classroom and revitalizing productivity.
- Exceeded facility's goals by training over 98% of squadron personnel.
- Trained all users and wrote over 100 pages of operating manual.
- Managed all phases of work order completion from receipt of order to finished product.

MILITARY ASSIGNMENTS

20__ - Present **15th Civil Engineering Squadron**, Pope AFB, NC

20__ -20__ **514th Civil Engineering Squadron**, Scott AFB, IL

SECURITY CLEARANCE

Top Secret with current Special Background Investigation

Resume provided by Rita Fisher, CPRW; RitaFisher33@comcast.net

Electrician

FRANCIS X. SEABEE

San Diego, CA
(h) 619-555-9999 (c) 410-666-1234
SeabeeF@gmail.com | linkedin.com/fseabee

OBJECTIVE Position as an electrician for a company seeking an experienced, skilled technician.

QUALIFICATIONS

- Six years experience as an Industrial Electrician for the U.S. Army
- Rated superior in all technical repair activities
- Demonstrated ability to maintain, operate, repair and install a wide variety of commercial and electronic equipment
- Secret security clearance with current special background investigation

SKILLS SUMMARY

Experienced in preventative and corrective electrical maintenance on the following equipment:

- 3 phase and single phase AC / DC plant equipment
- Power and lighting circuits, switches, and fuse boxes
- Amp meters, volt meters, ohm meters, and other test equipment
- Alarm systems and other power monitoring equipment

Knowledgeable at using the following tools to troubleshoot and repair equipment malfunctions:

- Logic test equipment
- Power presses
- Automatic testing
- Undercutting machines
- Spot welding machine
- Coil winding machine
- Trickle and impregnation machine
- Soldering tools

Experienced in reading blueprints/drawings and using the following tools to install or repair cables, conduits, and circuits:

- Conduit benders
- Hand and power tools
- Pipe threaders
- Cable pullers
- Wire and cable cutters

Supervised the operation of turbine generators and emergency diesel generators. Trained junior staff.

EMPLOYMENT TRAINING

Electrician 400 Hz Motor-Generator Maintenance, 20__

Electrician "C" School, 20__

Electrician "A" School, 20__

EDUCATION

Diploma, Edison High School, Edison, NJ 20__

Certificate (Honor Graduate), AH-64 Armament / Electrical System Repairer Course, Ft. Eustis, VA, 20__

Sales Trainer

Available: June 20__

JAMES L. PARK

Quantico, VA
W: (703) 222-3333 / H: (703) 444-5555
ParkJ@gmail.com
Linkedin.com/jameslpark

OBJECTIVE

Sales training position for a consumer products company seeking highly motivated individual with outstanding salesmanship skills.

QUALIFICATIONS SUMMARY

RECRUITING

- Interacted with candidates and their parents on a daily basis.
- Highlighted the benefi ts of military service.
- Surpassed recruitment goals by 15% annually.
- Visited over 100 local high schools and community colleges to attract top-notch Marine candidates.

TRAINING

- Indoctrinated 150 new recruits in the traditions and practices of the Marine Corps.
- Physically and mentally challenged new recruits, preparing them for a wide range of national security-related assignments.
- Trained new recruits on the use of various weapon systems.
- Achieved 95% unit proficiency.

MANAGEMENT

- Led and directed the activities of a 10-member team; improved both their individual and team skills.
- Counseled and mentored subordinates, evaluated their performance, and provided developmental advice.

EMPLOYMENT HISTORY

Recruiter, U.S. Marine Corps, Atlanta, GA, 20__-Present
Drill Instructor, U.S. Marine Corps, Paris Island, SC, 20__-20__
Squad Leader, U.S. Marine Corps, Okinawa, Japan, 20__-20__
Member of Squad, U.S. Marine Corps, Okinawa, Japan, 20__-20__

EDUCATION & TRAINING

A.S. Business Administration, Albany Community College, GA, 20__
U.S. Marine Corps Recruiters School, 20__
Advanced Leadership Training, 20__
U.S. Marine Corps Basic Training, 20__

Management - Security

THOMAS BARNES

Macon, GA * 478-742-2442 * thomasb@gmail.com * linkedin.com/thomasbarnes

SUMMARY OF QUALIFICATIONS

Training / Law Enforcement / Management

Dynamic Management, Training, and Law Enforcement Professional with more than 20 years of demonstrated achievements gained while serving in the United States Army. Hold Secret Security Clearance. Successfully managed and maintained more than $4M worth of harbor craft, police vehicles, and various law-enforcement equipment. Coordinated and supervised all equipment and personnel involved with complex training plan eventually adopted for use on all bases in Mid-Atlantic Region.

PROFICIENCIES SYNOPSIS

* Management	* Project Planning	* Supervision
* Training	* Mentoring	* Problem-Solving
* Curriculum Development	* Interpersonal Communication	* Law Enforcement

CAREER OVERVIEW AND HIGHLIGHTED ACCOMPLISHMENTS

PROGRAM MANAGER/LEAD INSTRUCTOR, U.S. Army, Fort Sill, OK — April 20__ – Present

Supervised the daily operations and well-being of 265 personnel. Chosen by unit Commander to supervise 45 personnel, 2 vehicles, and an engineering shop.

- Assisted in writing comprehensive regional Patrol Standard Operating Procedure (SOP).
- Interfaced and coordinated with diverse group of personnel at all levels of the organization.
- Personally developed training program adopted by Mid-Atlantic Region.

TRAINING DIVISION SUPERVISOR, U.S. ARMY, Fort Eustis, VA — April 20__ – April 20__

Served as lead driving instructor, providing instruction and qualifying more than 60 personnel. Managed the upkeep and maintenance of 20 vehicles worth an excess of $5.5M.

- Supervised and prioritized the daily work and operations of 30 personnel maintaining 100% safety.
- Managed 15 armory personnel, 289 small arms, and 1 mobile firing range.
- Created and implemented a comprehensive training program, including two weeks of classroom instruction.
- Handpicked by unit Commander to restructure and manage Precinct Division.

PATROLMAN, U.S. ARMY, Army Base Guam Security Department, Guam — Jan 20__ – March 20__

Helped to create a Harbor Police Division from the ground up, utilizing personnel and resources to optimal efficiency.

- Trained and qualified 45 personnel on numerous law-enforcement operating and safety procedures.
- Wrote and improved numerous SOPs and other official documentation and briefed leadership.
- Supervised and prioritized the work of 30 direct report personnel ensuring the safety and protection of a 10 acre facility.

EDUCATION AND CERTIFICATIONS

BA, Organizational Development, University of Maryland

Non-Lethal Weapons Course Instructor
Emergency Vehicle Operator Course Instructor
Army Small Arms Instructor Certification (Top in Class)

AWARD RECOGNITION

Meritorious Achievement Medal (2)
Army Commendation medal (3)
Humanitarian Service Medal

Resume provided by Barbara Adams at Career Pro Global; Barbara@careerprocenter.net

Executive - Intelligence

IAN MOORE
Springfield, VA, 703-555-7733
ianmoore@email.com | www.linkedin.com/ian-moore

DIRECTOR, BUSINESS INTELLIGENCE AND OPERATIONS

INTELLIGENCE ANALYSIS / OPERATIONS / PROGRAM MANAGEMENT

Strategic leader with a Top Secret/SCI Clearance and over 28 years of U.S. Army worldwide experience in Military Intelligence, Logistics and Diplomacy. As a Foreign Area Officer managed **international military and diplomatic programs, procurement processes, business development**. Led an International Coalition team and contractors in Afghanistan to **modernize Civilian Afghan Police intelligence operations and training** resulting in a 60% increase in police analysis, operations.

Recognized by Deputy Assistant Secretary of Defense as a "leader and dynamic global talent with exceptional negotiation and management skills."

CORE COMPETENCIES

- Policy & Planning
- Interagency Coordination
- Program Management
- International Diplomacy Analysis
- Strategic, Operational & Tactical Force
- Global All-Source Intelligence Analysis & Production

PROFESSIONAL EXPERIENCE

Military Faculty, National War College **Jun 20__ – Present**
National Defense University, Washington, DC
Instructor and Advisor educating future leaders of the Armed Forces, Department of State, International students and other civilian agencies for high-level policy, strategic planning, command responsibilities

- Taught senior-level curriculum in national security strategy and national security policy, resulting in a "first time" 100% pass rate.
- Recognized by the Chairman of the Joints Chief of Staff with the Defense Superior Service Medal for being "a driving force in training senior military officers and civilians throughout the interagency and across the globe."

Director, Initiatives Group, Office of the Deputy Chief of Staff **Jun 20__ – Jun 20__**
U.S. Army, G2 Intelligence, The Pentagon
Managed all operations and administration in support of the G-2's priorities. Provided analysis and planning interface across the HQDA staff for doctrinal Army planning, policy and programming for the resourcing and requirements process review for the U.S. Army worldwide.

- Led G2 strategic communications efforts ensuring executive communication met Deputy Chief's intent and vision.

Director and Sr. Advisor, Afghan Director of Police Intelligence **Jun 20__ – Jun 20__**
NATO Training Mission - Afghanistan
Primary Advisor to the Afghan Director of Police Intelligence (DPI) and Civilian Police - 2423 intelligence personnel.

- Supervised all force generation actions to develop the DPI to include manning, equipping, resourcing, training, building the National Technical Exploitation Center, and DPI Training Center.
- Supervised and led a section of over 275 U.S. Coalition soldiers and Contractors in combat operations.

Director, Joint Staff, J2, Directorate for Current Intelligence **Jul 20__ – Jul 20__**
The Pentagon
Directed the Intelligence Operations Division in the National Joint Operations and Intelligence Center (NJOIC), Pentagon. Coordinated the activities of CIA, NSA, NGA, NRO, Defense HUMINT, DoD Collections, and Military Component representatives. Monitored worldwide all-source intelligence reporting of potential threats to U.S. national interests.

- Single point of communication and force analysis to the Chad Defense Attaché and the U.S. Embassy during the U.S. Embassy collapse.

EDUCATION

MS, Nat'l Resource Mgmt., Eisenhower Industrial College of the Armed Forces, National Defense University
MS, Eurasian Studies, Indiana University
BA, Psychology, Wilkes University

Resume provided by Janet Spadola, The JAS GROUP INC, janet@thejasgroupinc.com, Transitioning Military & Executive Leaders

Aviation Logistics

Susan Brooks
Boston, MA
(617) 555-5555
Sbrooks70@gmail.com
Linkedin.com/susanbrooks

SUMMARY

Detail-oriented logistician with a focus on providing logistical support in inventory control, financial accountability as well as material procurement in the aviation industry. Continuously challenges and scrutinizes the supply system process by utilizing analytical skills to identify deficiencies and excess. Exceptional in interpreting and adhering to logistics principles and government standards outlined in established policies, regulation, and laws.

KEY QUALIFICATIONS

- Data Driven
- Attention to Detail
- Leadership
- Teamwork
- Top Secret Security Clearance
- Analytical and Critical Thinking
- Multi-tasking
- Adaptability
- Problem Solving
- Communication

EXPERIENCE

Material Support

- Supervised the procurement of over 2,540 routine requisitions for aircraft components valued in excess of $5.4M over a 17-month period.
- Expedited and oversaw the daily reporting of 820 outstanding critical components valued at $2.9M in support of 12 executive support aircraft contributing to an unprecedented daily readiness rating of 90%.
- Collaborated with the department's warehouse managers to improve stock posture from the established CH-46 to MV-22 helicopter platforms thereby continuously improving turnaround time.
- Managed a warehouse of 950 protective flight clothing equipment adhering to strict aviation personal safety standards for over 800 individuals.

Accountability

- Ensured 100% accountability of over 360 non-critical but essential support assets valued at more than $450,000 by performing regular inventories and meeting allowance objectives by procuring or properly disposing materials as needed.
- Meticulously monitored the safe disposition of over 400 pieces of material to the Defense Reutilization and Marketing Office (DRMO).
- Performed a variety of warehouse functions to maintain over 1,200 line items of repairable aircraft components worth over $200M supporting various aircraft platforms and during the upgrade of the H-1 helicopter.

Aviation Logistics - Page 2

Financial

- Oversaw and continually monitored the budget of 230,000 transactions for aircraft components totaling over $360M over 24 months.
- Consistently met all financial reporting deadlines by meticulously analyzing transaction data for discrepancies and performing corrective action impressing system integrity.
- Managed funding for the Government Commercial Purchase Card accounting for over $340,000 in purchases a year by judiciously reviewing all orders and ensuring strict adherence to the program's guidelines.

Communications

- Established rapport with inventory managers and external naval and defense activities to expedite critical or high demand components; relaying essential demand data to those activities resulting in more accurate forecasts for these components.
- Effectively communicated with leadership on requisition data, identifying areas for improvement and reporting critical technical information.

Leadership

- Mentored and counseled individuals to the successful execution of unit and base level recognition boards resulting in the professional achievement of 7 individuals.
- Planned and developed physical training sessions, professional education classes, and morale functions to promote an atmosphere of teamwork, ethical decision-making, and opportunities to exercise leadership at different levels.

EMPLOYMENT HISTORY

- Flight Detachment (USMC) – Camp Lejeune, North Carolina 20__-Present
- Marine Aviation Squadron – Twentynine Palms, California 20__-20__

EDUCATION

University of Nebraska – Lincoln, Nebraska 20__-20__
Bachelor of Science in Business

Mechanical Repair

Samuel Keller
skeller@gmail.com
Macon, GA
(478) 742-2442
Linkedin.com/samkeller

AREAS OF EXPERTISE:

- Electrical/Mechanical Systems
- Analysis and Troubleshooting
- Preventive Maintenance
- Leadership and Training
- Quality Control
- Organizational Management
- Communication and Technology

SUPERVISOR COMMENTS:

"Demonstrated versatility in all assigned endeavors. Directly responsible for meeting 100% of DC readiness."
--M.A. Feeney

"Highly motivated. Assumes and exhibits organizational abilities, technical skills, and greater responsibility..."
--J.W. Wilson

"Regularly demonstrates the ability and capacity to successfully assume positions of greater authority and responsibility."
--A.J. Andrews

ELECTRICAL – MECHANICAL SYSTEMS MAINTENANCE

Motivated professional offering a Secret Security Clearance with expertise in gas turbine engine systems, repair, troubleshooting, and preventive maintenance. Provides technical support and ensure readiness for over $4M in equipment assets. Strong leadership, training, administration, and mentoring attributes. Communicates clearly to define expectations and mentor subordinates, effectively identifying problems, providing intuitive analysis, and developing innovative solutions. Achieves excellence independently and as a team player. Thrives and excels in multitasking and prioritizing assignments in a fast-paced environment.

PROFESSIONAL EXPERIENCE

US Air Force Reserve, Peru, IN 03/20__ - Present
Jet Engine Mechanic

- Perform preventive maintenance and emergency repairs on KC135 jet aircraft engines.
- Maintain extensive knowledge of fuel, pumps, compressors, propulsion devices, and valves.
- Ensure material readiness valued at over $4M for all assigned systems
- Troubleshoot and restore equipment and systems to optimal working condition.
- Accomplished Jet Engine Mechanic earning recognition for mechanical maintenance skill.

US Navy, San Diego, CA 03/20__ – 04/20__
Training Specialist

- Supervised six divisional training officers in their duties and assignments.
- Completed the preparation, planning, and scheduling of Navy school requests, approving assignments of 54 military personnel and 17 civilian employees.
- Maintained high standards of performance as a quality assurance advisor.
- Managed $1M+ of assets, materials, and equipment with zero loss.

US Navy, Oahu, HI 02/20__ – 02/20__
Gas Turbine Technician/Main Propulsion

- Maintained marine propulsion engines and auxiliary equipment in optimal working condition.
- Performed valve maintenance and repaired hydraulic and pneumatic valves.
- Monitored and distributed all damage control equipment, including firefighting tools and training instruments, personally hydrotesting 30+ fire hoses for performance quality.
- Dispatched to emergency areas as a highly-skilled fireman and damage control specialist.
- Carried out technical support for all ships, submarines, aircraft carriers, and marine crafts.

EDUCATION and TRAINING:

Embry Riddle Satellite Courses, San Diego, CA – Associate's Degree, 31 semester hours (in progress); **Vincennes/Purdue Aviation Technology School**, Indianapolis, IN – 75 semester hours in power plant certification; US Air Force CFM56, TF33 Engine Course; Engineering Mechanical Core; Gas Turbine Mechanical/Electrical Course; Gas Turbine Console Operator; Mechanical Maintenance; Journeyman Instructor Training; First Line Leadership Development Program; Onboard Maintenance Training – Valve Repair.

AWARDS:

Armed Forces Expeditionary Medal (2); Sea Service Deployment Ribbon (3); Good Conduct Medal (4); Coast Guard Special Operations Ribbon; Navy and Marine Corps Achievement Medal; Navy "E" Ribbon; National Defense Service; Global War on Terrorism Expeditionary Medal; Flag Letter of Commendation.

Resume provided by Barbara Adams at Career Pro Global; Barbara@careerprocenter.net

COVER LETTER

615 Meadow Ridge Drive
Raleigh, North Carolina 27601
February 10, 20__

Robert Collins
Human Resources Director
ABC Company
480 K Street
Washington, D.C. 20001

Dear Mr. Collins,

I have applied online to the position of Program Manager, National Division at ABC Company that was posted on LinkedIn. I am confident that my skills are well aligned with this role and that I would be an excellent fit for your organization.

I am currently in the process of transitioning after ten years of military service and have recently achieved my PMP certification. I have demonstrated the ability to manage contracts from the proposal phase through implementation, with ever-increasing profitability. I have ideas for business development in diversified areas of finance and defense that could increase the size of your National Division.

I believe my military leadership experience, coupled with my education, make me an excellent fit for this position. I have enclosed my resume and would appreciate your giving me the opportunity to discuss my qualifications and ideas in person.

Sincerely,

Jonathan Hartman

Jonathan Hartman
jhartman@gmail.com
765-123-4567

COVER LETTER

321 Broad Street
Boston, MA 02123
November 4, 20__

Gerald Hastings
Director of Systems Development
Computer Systems, Inc.
4543 Computer Boulevard
Boston, MA 02112

Dear Mr. Hastings,

Please consider my application for the Systems Development Manager position at Computer Systems, Inc. that was posted on Corporate Gray Online. I believe that I possess the required skills and background as well as personal qualifications which would make me an asset to your company.

For the past five years I have had the opportunity to become an expert in software development life cycle and system requirements design, which I've truly enjoyed. I am looking forward to applying my knowledge and skills in a new setting with an established company where I can continue to grow in this field. In addition to my expertise in developing full life-cycle of systems, my fifteen years of experience in debugging and testing computer systems, coupled with my management experience, would make me a valuable asset to both your company and your clients.

After researching Computer Systems, Inc.'s vision/mission, I believe that my experience and career goals are an excellent match for your company and this position. I would like to discuss my qualifications in an interview at your convenience.

Best regards,

Laura Graystone

Laura Graystone
laurag@gmail.com
603-555-1234

APPROACH LETTER
Referral

555 Main Street
Fairfax, VA 22101
April 2, 20__

Abigail Carter
Senior Director of Engineering
ABC Company
627 Fulton Drive, Suite 201
Rockville, MD 20847

Dear Ms. Carter,

I am leaving the service after twelve years as a Systems Administrator for the U.S. Army. Bob Martin of your Engineering Division suggested I contact you in regards to civilian systems administration careers.

During my experience in the Army, I have held several leadership positions and have been instrumental in designing and maintaining critical military networks. Before undertaking my civilian job search, I am seeking advice from experienced professionals in the field to gain more knowledge of the civilian workplace.

I would appreciate the opportunity to meet with you briefly in the next two weeks to discuss my career plans and ask you questions about the industry. I will call your office on Tuesday, April 10th, to schedule a meeting.

I look forward to talking with you soon.

Sincerely yours,

James Smith

James Smith
jsmith@gmail.com
703-555-1234

THANK YOU LETTER
Post-Informational Interview

67221 Bayview Road
Jacksonville, FL 32208
October 2, 20__

Bernard Armstrong, Director
Southern Bank Services
835 Buckingham Boulevard
Atlanta, GA 30301

Dear Mr. Armstrong,

Thank you for meeting with me yesterday and helping me clarify my career goals in the finance industry. Your advice on certifications and experience was most helpful in guiding my decisions on which jobs to pursue. I appreciate you sharing your expertise with me and offering suggestions for improving my resume. I will incorporate your thoughtful ideas and send you an updated copy next week.

I have reached out to Mr. Haskins at Premier Financial Services as you suggested. I appreciate the referral and look forward to pursuing opportunities at his organization.

Thank you again for taking time from your busy schedule to meet with me. I will keep in touch and let you know how my job search progresses.

Sincerely yours,

Roger Dugan

Roger Dugan
dugan.roger@gmail.com
616-555-1234

THANK YOU LETTER
Post-Job Interview

11205 Ivey Court
Chesapeake, VA 23320
April 30, 20__

Marianne MacDonald
Director, Mid-Atlantic Division
Premier Services, Inc.
5779 Otis Avenue
Norfolk, VA 23510

Dear Ms. MacDonald,

Thank you for the opportunity to interview for the Program Manager position. I appreciate your hospitality and enjoyed meeting you and the other Mid-Atlantic Division group leaders.

As I mentioned during the interview, my experience and training in Program and Contract Management make me ideally suited to be a positive contributor to the Mid-Atlantic Division. I am confident I can help grow the division in the areas of finance and defense, which will result in expansion of services and increased profits.

Meeting with you and the other leaders convinced me that Premier Services, Inc. is an organization where I would enjoy working and where I could use my skills and experience to contribute to its future successes.

Please let me know if you need more information about my background or references. I look forward to hearing from you.

Sincerely yours,

Raymond Peterson

Raymond Peterson
rpeterson@gmail.com
757-123-4567

chapter 7

Conduct Research in Key Areas

THE OLD ADAGE THAT "knowledge is power" is especially true when conducting a job search. Your job search is only as good as the knowledge you acquire and use for finding the job you want. Gathering, processing, and using information is the lifeblood of any job search. Research integrates the individual job search activities and provides feedback for adapting strategies to the realities of the job market. Given the numerous individuals and organizations involved in your job search, you must develop an information gathering strategy that will help you gain knowledge about, as well as access to them.

Investigate Alternative Jobs and Careers

Your initial research should help familiarize you with job and career alternatives. For example, the U.S. Department of Labor identifies approximately 13,000 job titles. Most individuals are unaware of the vast array of available jobs and careers. Therefore, it is essential to investigate occupational alternatives to broaden your perspective on the job market. It is especially important for you to discover how your military job skills and titles best correspond to specific civilian job skills and titles. You might start by looking at your VMET document. After that, you should start your research by examining several key directories that provide information on alternative jobs and careers:

- *CareerOneStop Toolkit* (www.careeronestop.org/Toolkit)
- *The Occupational Outlook Handbook* (www.bls.gov/ooh)
- *The O*NET Occupation Search* (https://www.onetonline.org/find)

Conduct Industry and Company Research

Conducting online research will help to strengthen and clarify your career objective. There are many websites that function as excellent gateways, online databases, and research tools. As you research careers, identify specific organizations which you are interested in learning more about. The most important information you should be gathering concerns the organizations' goals, structures, functions, problems, and projected future opportunities and development. Since you will be investing part of your life in such organizations, treat them as you would a potential stock market investment. Compare and evaluate different organizations.

Research Tools

We recommend signing up for a **LinkedIn** account (www.linkedin.com) and using the social media site to research companies and careers. Using the search box, you can search by keyword for companies in your industry. Do an advanced search to filter your search further by location, size, etc. If you go to a company's LinkedIn page you'll see more information about the company and its employees.

Also use LinkedIn to research people. By doing a people search, you can study the profiles of others in your industry and see what keywords they are including, what education and skills they possess, who they've worked for in the past, and other pertinent data.

Other online tools to help you research industries and companies are:

- CEO Express www.ceoexpress.com
- Chamber of Commerce https://www.uschamber.com
- Corporate Information www.corporateinformation.com
- D&B's Million Dollar Database www.mergentmddi.com
- Forbes Lists www.forbes.com/lists
- Fortune 500 www.fortune.com
- GlassDoor www.glassdoor.com
- Hoover's Online www.hoovers.com
- Inc. 5000 www.inc.com/inc5000
- Manta Small Business Directory www.manta.com
- Moody"s www.moodys.com
- NASDAQ www.nasdaq.com
- Newspapers.com www.newspapers.com
- Salary.com www.salary.com
- S&P Global www.standardandpoors.com
- Thomas for Industry www.thomasnet.com

Directories of Career and Job Alternatives

- Career Outlook www.bls.gov/careeroutlook
- Occupational Outlook Handbook www.bls.gov/ooh

Industrial Directories

- Industrial Quick Search www.iqsdirectory.com
- Industrial Web Directory www.industrialwebdirectory.com

Associations

- Directory of Associations www.directoryofassociations.com

Government Sources

- GovEngine Federal Government List https://govengine.com/fedgov
- Index of U.S. Government Agencies www.usa.gov/federal-agencies
- Congressional Directory www.govinfo.gov/app/collection/cdir

Newspapers

- Major city newspapers and trade newspapers. Many are available online through these major gateway sites: www.onlinenewspapers.com and www.newspapers.com.
- Your targeted city newspaper—the Sunday edition.

Business Publications

- *Business Week, Economist, Fast Company, Forbes, Fortune, Harvard Business Review, Newsweek, Red Herring, Smart Money, Time, U.S. News and World Report.* Many of these and other business-oriented publications can be viewed online through www.CEOExpress.com.
- Annual issues of publications surveying the best jobs and employers for the year: *Money*, *Fortune*, *Forbes*, and *U.S. News and World Report*. Several of these reports and publications are available online: www.cnn.com/business, www.fortune.com, and www.forbes.com/lists.

Other Useful Resources

- Trade journals (www.webwire.com/IndustryList.asp)
- Publications of Chambers of Commerce, state manufacturing associations, and federal, state, and local government agencies

Contact Individuals

While examining directories and searching the Internet for alternative careers, industries and companies will provide you with useful job search information, much of this material may be too general for specifying the right job for you. In the end, the best information will come directly from knowledgeable people in your targeted organizations.

As you research companies, start compiling lists of names and contact information for important individuals in each organization. Your most productive research activity will be communicating with them by email, LinkedIn, and phone. You especially want to learn more about the people who make the hiring decisions. To help you stay organized, you can use the Company Contacts worksheet, or similar tool, found under Step 7 of the Online Transition Guide at www.corporategray.com/static/transition_guide/conduct_company_research.

Conducting **informational interviews** may be the most useful way to research careers, organizations, and job opportunities. An informational interview is an informal conversation that you set up to ask advice and get information from professionals in your targeted industry. This procedure is described in detail in Chapter 3.

Use the following websites to connect with former military friends and colleagues who are in the civilian work world and may even be hiring managers for their companies:

- Facebook (www.facebook.com)
- LinkedIn (www.LinkedIn.com)
- Vet Friends (www.vetfriends.com)

Ask the Right Questions

The quality of your research will only be as good as the questions you ask. Therefore, you should focus on a few key questions that will yield useful information for guiding your job search. Answers to these questions will help make important job search decisions relevant to informational and job interviews.

"Who Has the Power to Hire?"

Finding out who has the power to hire may take some research effort on your part. Keep in mind that Human Resources offices normally do not have the power to hire. They handle much of the paperwork and perform auxiliary support functions for those who do the hiring—usually individuals in the operating business units.

If you want to learn who really has the power to hire, you need to conduct research on the particular operating unit that interests you. You should ask specific questions concerning who normally is responsible for various parts of the hiring process:

- Who describes the positions?
- Who announces vacancies?
- Who receives applications?
- Who administers tests?
- Who selects eligible candidates?
- Who interviews the candidates?
- Who makes the hiring decision?

If you ask these questions about a specific position you will quickly identify who has what powers to hire. Chances are the power to hire is shared between the Human Resources office and the operating unit. You should not neglect the Human Resources office, and in some cases it will play a powerful role in all aspects of the hiring. Your research will reveal to what degree the hiring function has been centralized, decentralized, or fragmented.

"How Does Organization X Operate?"

It's best to know something about the internal operation of an organization before joining it. But the type of knowledge we recommend you seek goes far beyond the structure of the company or knowledge about its products and services. Yes, these are important and we encourage you to know them. However, the more challenging aspect of your research is to learn about the organization's **corporate culture**, which refers to the overall environment or climate that exists within an organization.

Some organizations are high-pressure, where there is an ever present demand to increase the pace of business and profitability. Other organizations might be more laid-back, where the emphasis is on refining the quality of a product or service rather than the amount of business each person is generating. A company's corporate culture permeates the organization. It reflects the values and morals of the organization's senior leaders.

In many ways you are already familiar with a corporate culture—your service's. Your research into prospective companies should help you determine whether you'll fit in to their corporate culture. When possible, we suggest trying to solicit this information in your informational interviews.

There are many other, more pragmatic reasons for conducting a thorough research of prospective companies. Your research may uncover information that would convince you that a particular organization is not one in which you wish to invest your time. You can get financial information about most companies by examining their annual reports and by talking to individuals who know the organization well. To obtain a copy of a publicly owned company's annual report, simply call and ask. Most will give you a free copy. If, on the other hand, the company is privately owned, they will probably decline your request.

With regard to information on internal operations, especially company politics and power, you must glean this information from individuals who work within the organization. Ask them: "Is this a good organization to work for?" and let them expand on specific areas you wish to probe—advancement opportunities, working conditions, relationships among co-workers and supervisors, growth patterns, internal politics, management styles, work values, or opportunities for taking initiative.

"What Do I Need to Do to Get a Job With Organization X?"

The best way to find how to get a job in a particular organization is to follow the advice in Chapter 3 on prospecting, networking, and informational interviewing. This question can only be answered by talking to people who know formal and informal hiring practices.

You can get information on a company's hiring process by contacting their Human Resources office. But you must go beyond the formal system and Human Resources office in order to learn how best to conduct your job search. This means contacting people who know how one really gets hired in the organization. The best sources of information will be individuals who play a major role in the hiring process.

Identify the Right Community

As you separate or retire from the service, identifying the geographical area where you would like to work will be one of your most important career decisions. Once you make this decision, other job search decisions and activities become easier.

There are important considerations in choosing the area that best fits you and your family. You should evaluate the educational and employment opportunities, quality of schools, community feel and cultural activities, among other aspects that may be of importance to you. To help you explore various communities, examine several of these gateway community sites:

- AreaVibes www.areavibes.com
- Boulevards https://blvds.com
- Chambers of Commerce www.chamberofcommerce.com
- City-Data www.city-data.com
- Sperling's Best Places www.bestplaces.net

For further assistance in conducting company research and for a checklist of links to online resources listed in this chapter, visit www.CorporateGray.com > Transition Guide > Conduct Company Research.

STOP
Do not close this book until you create a free account on Corporate Gray Online.
Where transitioning service members meet Corporate America
Take 5 minutes. Create your free account now.
www.CorporateGray.com

chapter 8

Interview for the Right Job

MAKE NO MISTAKE—the job interview is the most important step in the job search process. All previous job search activities lead to this one. It is important that you interview for the right job. By this we mean one that is compatible with your knowledge, skills and interests. Only you can judge what constitutes the right job. You want to find a job that you will enjoy and do well. Your previous job search activities—skills assessment, research, and informational interviews—should provide you with a strong indication as to whether a given job is right for you. Once you've done your homework and found an employment opportunity that appears right for you, it's time to prepare for the job interview. As we've stressed in previous chapters, proper preparation is key. How you approach the interview will make a difference in the outcome of the interview. Therefore, you need to know what best to do and not to do in order to make an excellent impression on employers.

Interviewing For the Job

While the job interview is the most important job search activity, it also is the most stressful job search experience. Your application, resume, and letters may get you to the interview, but you must perform well in person in order to get a job offer. Knowing the stakes are high, most people face interviews with dry throats and sweaty palms. But careful preparation will help reduce the stress level.

How do you prepare for the interview? First, you need to understand the nature and purpose of the interview. Practice the entire interviewing scenario, from the time you enter the door until you leave. You should sharpen your nonverbal communication skills and be prepared to give positive answers to questions as well as ask intelligent questions. The more you practice, the better prepared you will be for the real job interview.

Answering Questions

Once you receive an invitation to interview, prepare as if it were a $1,000,000 prize. After all, that may be what you earn during your employment.

Phone Interviews

In today's competitive job market, hiring managers often receive hundreds of resumes. To make best use of their time, most hiring managers will filter through the stack of resumes and select the most promising ones for a phone screen. This telephone interview is an important early discriminator, so you must be prepared for it. Knowing that phone screens are important, you can gain a competitive advantage over less motivated job seekers through careful preparation. We recommend you start by developing a telephone script. A script is a series of questions you expect to be asked along with notes on your prepared responses. The following are examples of questions you might encounter:

- What type of work are you looking for?
- What do you consider your strongest skills? Weakest skills?
- How did you hear about this job opportunity?
- What do you know about our company and why are you interested in working for us?
- What level of education/training do you have?
- What are you doing now?
- Are you willing to relocate?
- Are you willing to travel? How much (what percentage)?
- When would you be available to start work?
- When could you meet with us?

In response to each question, you should develop a clear, concise answer. By developing and practicing a script, you'll be ready to give polished answers in a confident tone. When you're done with your preparations, keep a copy handy so you'll be ready when the hiring manager calls. Remember, the phone screen is often when the prospective employer will get his first impression of you. Your job is to make it as positive as possible.

Video Interviews

Virtual video interviews are an increasingly important part of the job search process. Using this technology, companies can more easily screen candidates regardless of their geographic location, which is especially useful for military personnel who are stationed around the globe.

These video interviews allow for a virtual face-to-face conversation between a recruiter and candidate. Some companies also use one-way video interviews, whereby the candidate answers a set of pre-screening questions that are recorded and viewed later by the recruiter.

Here are some video-specific tips to keep in mind:

- Dress appropriately from head-to-toe. Avoid patterned or white clothing.
- Choose a quiet place with a clean desk and simple background.
- Make sure your microphone and camera are adjusted properly.
- Close other programs on your computer and use the latest version of the browser.
- Take advantage of practice questions, if available.
- Look into the camera not at the screen.
- Know how much time is remaining so that your answer is not cut short.
- Don't be afraid to make a mistake and start your answer over.
- Avoid using hand movements, which are more distracting on video than in-person.
- Use the notes you prepared, but don't rely too heavily on them.

In-Person Interviews

Successful screening interviews often result in an in-person interview. Once you confirm an interview, thoroughly research the company and interviewer. Lessen your anxiety by practicing the interview situation. Preparation and practice are the keys to doing your best.

During the interview, you want to impress the interviewer by providing concise answers that relate your skills and experience to their needs. Where possible, your answers should incorporate your knowledge of the firm based on your research and networking activities. In so doing, you will help the interviewer see you as the right person for the job.

You should practice for the interview by addressing the questions interviewers are likely to ask. Most of these questions will relate to your education, work experience, career goals, personality, and related concerns. Some of the most frequently asked questions include:

Education

- Describe your educational background.
- How have you improved your education while in the military?
- Have you started work on an associate's or bachelor's degree? If not, why not?
- What military training courses did you take and how are they relevant to this job?
- Why did you attend ____________ (College or School)?
- What was your grade point average?
- How did you finance your education?
- If you started all over, what would you change about your education?
- What plans do you have to continue your education?
- What skills do you hope to acquire through education during the next 5 years?

Work Experience

- How many different jobs have you held?
- What were your major achievements in each job?
- What did you do in the military? Tell me about your different jobs.
- How does your military experience relate to this job?
- What did you enjoy the most about your military career? The least?
- What is your typical work day like?
- What functions do you enjoy doing the most?
- What did you like about your boss? Dislike?
- Which job did you enjoy the most/the least? Why?
- Have you ever been fired? Why?

Career Goals

- Why did you decide to leave the military?
- Why do you want to join our organization?
- Why do you think you are qualified for this position?
- What ideally would you like to do?
- Why should we hire you?
- How would you improve our operations?
- What do you want to be doing five years from now?
- What are your short-range and long-range career goals?
- If you could choose a job and organization, where would you go?
- What other jobs and companies are you considering?
- When will you be ready to begin work?
- How do you feel about relocating? Traveling? Working overtime?
- What attracted you to our organization?

Personality and Other Concerns

- Tell me about yourself.
- What are your major weaknesses? Your major strengths?
- What do you do in your spare time? Any hobbies?
- What types of books do you read?
- What role does your family play in your career?
- How well do you work under pressure? In meeting deadlines?
- Tell me about your management philosophy.
- How much initiative do you take?
- What types of people do you prefer working with?
- How (creative, analytical, tactful, etc.) are you?
- If you could change your life, what would you do differently?

Handle Objections and Negatives With Ease

Interviewers often have a healthy skepticism of job candidates, as they have encountered job seekers who have overstated their competencies and accomplishments. Although they do not always ask you these questions, they think about them nonetheless:

- Why should I hire you?
- What do you really want?
- What can you really do for me?
- What are your weaknesses?
- What problems will I have with you?
- How long will you stay?

Some employers may have objections to hiring people coming from a different organizational culture. How can you best handle employers' objections? You must first recognize their biases and stereotypes and then raise their expectations. You do this by stressing your strengths and avoiding your weaknesses.

When answering questions, both the substance and form of your answers should be positive. Compare your reactions to the following interview answers:

QUESTION: Why do you want to leave your present job?

ANSWER 1: After serving in the military for 20 years, I'm burned out. Morale isn't good and promotions are slow.

ANSWER 2: After serving in the military for 20 years, I've learned a great deal about leadership and teamwork. I have been given considerable responsibility for someone my age and demonstrated that I could handle it, even under high-stress situations. While I'm very proud of my military service, it's time to return to the civilian world and apply my skills and experience in that environment. I'm ready to take on more responsibilities as part of my professional growth.

The first answer communicates too many negatives. The second answer is positive and upbeat in its orientation toward skills, accomplishments, and the future.

In addition to choosing positive words, select content information which is positive and adds to the interviewer's knowledge about you. Avoid simplistic "yes/no" answers. Provide information which explains your reasons and motivations behind specific events or activities. For example, how do you react to these factual answers?

QUESTION: You have served in the military over seven years. As the government is a service-oriented organization, how will you fit into our profit-oriented environment?

ANSWER 1: I can understand your concern but feel I can learn to operate in a profit-oriented environment.

ANSWER 2: While working in the military is a service-oriented occupation, over the last few years the budget has been cut significantly and we have been forced to do "more with less." This pressure to become more cost effective has caused the military to find better and less expensive ways of doing business. I have done my part by recommending various cost-saving measures which have been implemented. For example, in the past we discarded brass casings of spent bullets at the firing range. I recommended that we collect these casings, melt them down, and reuse them in the manufacture of new bullets. The military accepted my idea, which ultimately saved the government over $100,000 per year. If I were given the opportunity to work for your firm, I would attempt to identify analogous cost saving measures that would improve your profitability.

The first answer misses an important opportunity to give evidence that you have resolved this issue in a positive manner. The second answer provides the interviewer with a rebuttal that tactfully diffuses his assertion using a specific example.

The most difficult challenge to your positive strategy comes when the interviewer asks you to describe your weaknesses:

QUESTION: We all have our weaknesses. What are yours?

This is not the time to bluntly confess your weaknesses. But you need to provide a thoughtful and intelligent response. In so doing, you can handle this question in any of five different ways, yet still give positive information on yourself:

1. **Discuss a negative not related to the job being considered:**

   I don't enjoy accounting. I know it's important, but I find it boring. Marketing is what I like to do. Other people are much better at bookkeeping than I am. I'm glad this job doesn't involve any accounting!

2. **Discuss a negative which the interviewer already knows:**

   Since graduating from high school, I have been serving in the military. Over the last couple years, I have been taking courses leading to a bachelor's degree. As a result, I currently lack civilian work experience. However, I believe my military training and the degree I will finish this year have prepared me well for this job. My leadership experience in the military taught me how to work with people, organize, and solve problems.

3. **Discuss a negative which you have improved upon:**

   I used to get overcommitted and miss important deadlines. But then I read a book on time management and learned what I was doing wrong. Within three weeks I reorganized my use of time and found I could meet my deadlines with little difficulty. The quality of my work also improved. Now I have time to work out at the gym each day. I'm doing more and feeling better at the same time.

4. **Discuss a negative which can also be a positive:**

   I'm somewhat of a workaholic. I'm now learning to more effectively manage my time so that I can achieve a better work-life balance. The results to date have been positive.

All of these examples stress the basic point about effective interviewing. Your single best strategy for managing the interview is to emphasize your strengths and positives. Questions come in several forms. Anticipate these questions, especially the negative ones, and practice positive responses in order to project your best self in an interview situation.

Encountering Behavior-Based Interviews

More and more employers are conducting behavior-based interviews, which are filled with behavior-related questions designed to elicit patterns of accomplishments relevant to the situation. They challenge interviewees to provide concrete examples of their achievements in different types of situations. Such interviews are based on the simple belief that how a job candidate has responded to certain types of situations in the past is a good predictor of how that person will behave in a similar future situation. Behavior-based questions are likely to begin with some variation of:

- Give me an example of a time when you . . .
- Give me an example of how you . . .
- Tell me what you did when . . .

This is an opportunity for you to sell your positives with an example or two. Briefly describe the situation, enthusiastically explain what you did (adding information as to why) and indicate the outcome.

You want to select examples that promote your skills and have a positive outcome. Even if the interviewer asks about a time when something negative happened, try to select an example where you were able to turn the situation around and get a positive result. If asked, "Tell me about a time you made a bad decision," try to identify an example where:

- Even though it wasn't the best decision, you were able to pull something positive out of the situation.
- Though a poor decision, you learned from it and in the next similar situation you made a good decision, or you know how you will handle it differently in the future.
- It was bad decision, but the negative outcome had only minor impact.

As you prepare for your interview, consider situations where you:

- demonstrated leadership
- solved a problem
- increased company profits
- made a good decision/made a poor decision
- handled change or trends
- handled criticism
- met a deadline/missed a deadline
- worked as part of a team

Add to this list other behavioral questions that apply to the job for which you are interviewing. For example, if the job includes making presentations, expect questions about a speech where you achieved your goal or conversely about a time when your speech failed.

Illegal Questions

Certain questions are illegal, but some employers ask them nonetheless. Consider how you would respond to these questions:

- Are you married, divorced, or single?
- How old are you?
- Do you go to church regularly?
- What does your spouse think of your career?
- What child care arrangements do you have?
- What organizations do you belong to?
- Do you own or rent your home?
- How much do you weigh?
- Do you have any disabilities?
- Are you in debt?

Don't get upset and say "That's an illegal question...I refuse to answer it!" While you may be perfectly right in saying so, this response lacks tact. If, for example, you are divorced and the interviewer asks about your divorce, you might respond with "Does a divorce have a direct bearing on the responsibilities of this position?" Some employers may ask such questions just to see how you react under stress. Others may do so out of ignorance of the law. Whatever the case, be prepared to tactfully handle these questions.

Asking Questions

Interviewers expect candidates to ask intelligent, thoughtful questions concerning the organization and the nature of the work. The nature and quality of your questions reveals your interest in the company and the job opening. Try to avoid asking self-centered questions that indicate you are primarily interested in knowing about salaries, benefits, perks, and advancement opportunities. Keep your questions employer- and job-centered. Consider asking some of these questions if they haven't been answered earlier in the interview:

- Please tell me about the duties and responsibilities of this job.
- How long has this position been open?
- How does this position relate to other positions in the company?
- What would be the ideal type of person for this position? Skills? Personality?
- With whom would I be working in this position?
- Who would be my first and second level managers?
- Please tell me something about these people? Their positions?
- What am I expected to accomplish during the first year?
- How will I be evaluated?
- What is the normal salary range for such a position?
- Being new, what problems might I initially encounter?
- Tell me about your experience here.
- What is particularly unique about working in this organization?
- What does the future look like for this company?

While it's better to internalize your questions, you may need to refer to your notes. You might do this by saying, "Yes, I jotted down a few questions which I want to make sure I ask you before leaving." Or, better yet, ask questions that have been triggered by the interviewer's earlier comments, and incorporate your research-based knowledge of the company. This will demonstrate that you're a good listener, are quick on your feet, and cared enough about the interview to learn more about a potential employer.

Dress Appropriately

Before you have a chance to speak, interviewers notice how you dress and accordingly draw certain conclusions about your personality and competence. Research shows that appearance makes the greatest difference when an evaluator has little information about the other person.

What you wear depends on the nature of the job. If you are pursing a white-collar position, select a classic-style suit in a conservative color, such as navy blue or charcoal gray. The best suit fabrics are wool or wool blends. For men, wear a long-sleeve, white cotton shirt with a point collar and a conservative tie. Women's suits also should be made of a natural fiber or have the "look" of a natural fiber. For the warmer climates or the summer months, linen blended with a synthetic or a good silk or silk blend are good choices.

When deciding on your professional wardrobe, always buy clothes to last, and buy quality. Men should look for suit jackets that are fully lined and pants that are lined to the knee. For women, silk blouses or a fabric that has the look and feel of silk are the fabrics to go with your suits. Choose your blouses in your most flattering shades. You can check out videos and images for proper interview dress by going to www.youtube.com or www.pinterest.com and searching under the key words "dress job interview."

Appear Likable

Remember, most people invited to a job interview have already been "screened in" and possess the basic qualifications for the job. In the end, employers hire people they like and who they believe will interact well on an interpersonal basis. Therefore, you should communicate that you are a likable candidate by engaging in several nonverbal behaviors. Four of the most important ones include:

1. **Sit with a very slight forward lean toward the interviewer.** It should be so slight as to be almost imperceptible. If not overdone, it communicates your interest in what the interviewer is saying.
2. **Make eye contact frequently, but don't overdo it.** Good eye contact establishes rapport with the interviewer.
3. **A moderate amount of smiling will also help reinforce your positive image.** You should smile enough to convey your positive attitude, but not so much that you will not be taken seriously.
4. **Try to convey interest and enthusiasm through your vocal inflections.** Your tone of voice can say a lot about you and your interest in the company.

Close the Interview

Be prepared to end the interview. Many people don't know when or how to close interviews. Interviewers normally will initiate the close by standing, shaking hands, and thanking you for coming to the interview. You should summarize the interview in terms of your interests, strengths, and goals. Briefly restate your qualifications and continued interest in working with the employer. At this point it is proper to ask the interviewer about selection plans: "*When do you anticipate making your final decision?*" Follow this question with your final one: "*May I call you next week to inquire about my status?*"

By taking the initiative in this manner, the employer will be prompted to clarify your status soon, and you will have an opportunity to talk further.

Many interviewers will ask you for a list of references. Be sure to prepare such a list prior to the interview. Include the names, phone numbers, and email addresses of four individuals who will give you positive professional and personal recommendations.

Remember to Follow Up

Once you have been interviewed, be sure to follow through to get nearer to the job offer. One of the best follow-up methods is the thank-you letter or email; you will find examples at the end of Chapter 6. In this correspondence express your gratitude for the opportunity to interview. Restate your interest in the position and highlight any particularly noteworthy points made in your conversation or anything you wish to further clarify. Close by mentioning that you will contact them in a few days to inquire about their decision.

A good way to keep track of your interviews is to maintain a record of each interview, such as with an Excel spreadsheet. You might include columns such as Company Name, Interviewer's Name, Interviewer's Phone Number, Interviewer's Email Address, Date of Interview, Interview Outcome, and Next Steps.

For more information and exercises to help you interview like a pro, visit www.CorporateGray. com > Transition Guide > Interview Like a Pro.

chapter 9

Negotiate Salary and Benefits

UP TO THIS POINT you have assessed your skills, conducted company research, interviewed for various opportunities, and are now preparing to negotiate salary and benefits with the company that is the best match for you. There are several questions that you need to consider. What are your salary requirements, and upon what are they based? What is the typical salary range for that occupation and location? What is your total military compensation? Do you know how to negotiate salary?

For most people transitioning from the military, these are difficult questions that require considerable thought and research. Since your entry on active duty, you've been accustomed to well-defined pay scales that dictate how much you'll be paid based on your pay grade, years of service, and marital status. That situation is about to change dramatically. After impressing upon the employer that you are the right person for the job, the bottom line becomes money—your talent and labor in exchange for the employer's cash and benefits. How, then, will you deal with these questions?

You may think you are worth a lot in the civilian work world—more than you have been getting paid in the military. But when it comes to questions of compensation, you must go beyond wishful thinking. As we will see in this chapter, you need to know how to value both the job and your skills and then translate these values into dollars.

Approach Salaries as Negotiable

Your military salary has been based on your rank and years of service. The situation is far different in the private sector. You must be prepared to negotiate your compensation based on your projected value to the employer.

Salary is an important consideration in the job search, but many individuals leaving the military are unprepared to properly handle the salary question, believing it is predetermined. However, most employers have some flexibility to negotiate salary.

Salaries are usually assigned to jobs rather than to individuals. But not everyone is of equal value in performing the job. Since individual performance differs, you should attempt to establish your value in the eyes of the employer rather than accept a salary figure for the job. The art of salary negotiation will help you do this.

Look to Your Financial Future

We all have financial needs which our salary helps to meet. But salary has other significance, too. It is an indicator of our worth to others. It also influences our future income. Therefore, it should be treated as one of the most serious considerations in the job interview.

The salary you receive today will influence your future earnings. Yearly salary increments will most likely be figured as a percentage of your base salary rather than reflect your actual job performance. Expect employers to offer you a salary similar to the one you earned in your last job. Once they learn what you made in your previous job, they will probably offer you no more than a 10% to 15% increase. If you hope to improve your income in the long run, then you must be willing to negotiate your salary from a position of strength.

Military Pay and Civilian Salary Parity

> The salary you receive today will influence your future earnings.

Many transitioning service members often undervalue themselves in the civilian work world because they tend to equate salary with base pay. If you've received base housing or a housing allowance, you know this benefit can be considerable. It translates into a specific dollar figure which should be added to your base salary. For example, if your base pay is $60,000 a year, you should include another 20 percent in benefits to arrive at a total compensation figure that would be equivalent to a civilian salary—around $72,000. If you only use your base pay as your current salary figure, you may undervalue yourself to civilian employers.

To determine your equivalent civilian compensation, visit http://paycharts.militarytimes.com.

Determine the Salary Range

There are multiple factors to consider when developing your salary range. Start with your equivalent civilian compensation as determined above. Also consider the research you've conducted thus far. Additional resources for obtaining salary information include:

- **Online salary surveys and employment sites:** Several websites include salary information. Among the most useful are www.Salary.com, www.SalaryExpert.com, www.PayScale.com, www.GlassDoor.com, www.SimplyHired.com/salaries, www.indeed.com/salary, and www.Vault.com.
- **Department of Labor's Bureau of Labor Statistics:** The BLS website gives an overview of wage data by area and occupation at www.bls.gov/bls/blswage.htm.
- **Trade and professional associations:** Most associations conduct annual salary surveys of their members. Contact them for the latest information on salary ranges.
- **Newspapers and print media:** Many of the larger newspapers provide online access to their classified section, which contains listings of job openings.

When computing your salary range, make the lower end of the range your minimum salary requirement. Then add 15 to 20 percent to that figure as the upper range.

Keep Salary Issues to the Very End

The question of salary may be raised anytime during the job search. Employers may want you to state a salary expectation figure early in the process. If at all possible, keep the salary question open until the end. Remember, you always talk about your TOTAL military compensation rather than your base pay. Even with application forms, cover letters, and telephone screening interviews, try to delay the discussion of salary by stating "open" or "negotiable." You should not attempt to translate your value into dollar figures until you have had a chance to convince the employer of your worth. This is best done near the end of the job interview process.

You should never ask about salary prior to being offered the job. Let the employer initiate the salary question. When he or she does, try to get the employer to state a figure first. If you do this, you will be in a stronger negotiating position. Negotiate the salary based on the research you have done in formulating your salary range.

Handle the Salary Question With Tact

When the salary question arises, your first step is to clearly summarize the job responsibilities/duties as you understand them. At this point you are attempting to do three things:

1. Seek clarification from the interviewer as to the actual job and all it involves.
2. Emphasize the level of skills required in the most positive way. In other words, you emphasize the value and worth of this position to the organization. This may help support the actual figure that the interviewer or you later provide.
3. Focus attention on your value in relation to the requirements of the position—the critical linkage for negotiating salary from a position of strength.

You might do this, for example, by saying:

As I understand it, I would report directly to the vice-president in charge of marketing and I would have full authority for marketing decisions that involved expenditures of up to $50,000. I would have a staff of five people—an administrative assistant, two copywriters, and two marketing assistants.

Although you may not explicitly draw the connection, you are emphasizing the value of this position to the organization. This position should be worth a lot more than one in which the hiree will report to the marketing manager, be required to get approval for all expenditures over $100, and has no staff. By doing this you will focus the salary question around the work you would be expected to perform in exchange for salary and benefits.

The next question to pose is:

What is your normal salary range for this position?

Now that the value and range for the position have been established, you can follow up with one more question.

What would be the normal salary range for someone with my qualifications?

This question further establishes the value for you rather than the position. This line of questioning will yield the salary expectations of the employer without revealing your desired salary figure or range. It also will indicate whether the employer distinguishes between individuals and positions when establishing salary figures.

Reach Common Ground and Agreement

After determining what the employer is prepared to offer, you have several choices. First, you can indicate that the figure is acceptable and conclude the negotiation process. Second, you can negotiate for greater compensation. Third, you can delay final action by asking for more time to consider the figure. Or, you can tell the employer the figure is unacceptable and leave.

The first and the last options indicate you are either too eager or playing hard-to-get. We recommend the second and third options. If you decide to reach agreement on salary in this interview, negotiate in a professional manner. You can do this best by establishing a salary range from which to bargain in relation to the employer's salary range. For example, if the employer indicates that he or she is prepared to offer $40,000 to $45,000, you should establish common ground for negotiation by placing your salary range into the employer's range. Your response to the employer's $40,000 to $45,000 range might be:

> *Yes, that does come near what I was expecting. I was thinking more in terms of $45,000 to $50,000.*

You, in effect, place the top of the employer's range into the bottom of your range. At this point you may be able to negotiate a salary of $42,000 to $43,000, depending on how much flexibility the employer has with salaries. Most employers have more flexibility than they admit.

Once you have placed your expectations at the top of the employer's salary range, you need to emphasize your value with supports, such as examples, illustrations, descriptions, definitions, statistics, comparisons, or testimonials. It is not enough to simply state you were "thinking" in a certain range; you must state why you believe you are worth what you want. Using statistics and comparisons as your supports, you might say, for example:

> *The salary surveys I have studied indicate that for the position of ________ in this industry and region the salary is between $65,000 and $70,000. I have extensive experience in all the areas you outlined. Taking everything in consideration—especially my skills and experience and what I see as my future contributions —I feel a salary of $70,000 is fair compensation. Is this possible?*

Another option is to ask the employer for time to think about the salary offer. You want to sleep on it for a day or two. A common courtesy is to give you at least 48 hours to consider an offer. During this time, you may want to carefully examine the job. Is it worth what you are being offered? Can you do better? What are other employers offering for comparable positions?

If you have other offers, this should put you in a stronger bargaining position. As a courtesy, contact the other employers. Let them know you have a job offer in hand, and give them the opportunity to make an offer of their own. We recommend that you carefully compare all job offers before making your decision. You will find a useful Job Offer Evaluation form at the end of this chapter, or download it from the Corporate Gray website in the Transition Guide > Step 9 - Negotiate Your Best Salary.

Carefully Examine Benefits

While serving in the military, you may have taken certain benefits, such as medical care, life insurance, and even physical fitness centers, for granted. In the civilian work world, these items and others like them have a direct cost that is carefully managed by employers. Given that the range of benefits may vary considerably from one company to the next, we recommend that you take these benefits into account when evaluating your job offers. Most companies will include some type of retirement savings plan (usually a 401k), health insurance, life insurance, short-term disability insurance, and paid vacations.

Offer a Renegotiation Option

You should make sure your future salary reflects your value. One approach to doing this is to reach an agreement to renegotiate your salary at a later date, perhaps in six months. Use this technique when you feel the final salary offer is less than what you are worth, but you want to accept the job. Be sure to get documentation of their willingness to renegotiate your salary at that later point.

Renegotiation provisions stress one very important point—you want to be paid on the basis of your performance. You demonstrate your professionalism, self-confidence, and competence by negotiating in this manner. More importantly, you ensure that the question of your monetary value will not be closed in the future. As you negotiate the present, you also negotiate your future with this as well as other employers.

Accepting the Job Offer

You should accept an offer only after reaching a salary agreement. Be sure to get the job offer in writing. Take time to consider your options and think carefully about the offer before giving the employer an answer. Accepting a job is serious business and you want to make the right decision.

Your job search is not over with the job offer and acceptance. You need to set the stage for launching a successful career with your new employer. Be thoughtful by sending your new employer a nice thank you letter. As discussed in Chapter 6, this is one of the most effective letters to write for getting your new job off on the right foot. The employer will remember you as a thoughtful individual and will look forward to working with you. The whole point of our job search methods is to clearly communicate to employers that you are competent and worthy of your salary.

To download the Job Offer Evaluation form and more information to help you negotiate your best salary, visit www.CorporateGray.com > Transition Guide > Negotiate Your Best Salary.

Job Offer Evaluation

Job Offer: ____

(Job Title, Company)

Scoring: 1 = Poor, 2 = Below Average, 3 = Average, 4 = Above Average, 5 = Excellent

Evaluation Factors	**Score**					**Notes**
	1	**2**	**3**	**4**	**5**	
Nature of Work						
Advancement Opportunity						
Training						
Work Environment						
Direct Supervisor						
Co-Workers						
Location						
Commute Time						
Salary						
Benefits (healthcare, insurance, etc.)						
Total Score:						

chapter 10

Secure Federal Employment

GOVERNMENT IS THE SINGLE LARGEST EMPLOYER in the United States, supporting more than 20 million employees. Federal, state, and local government agencies offer numerous attractive opportunities for transitioning service members. Federal government agencies, which employ 2.8 million civilian workers, are of special interest to individuals with military backgrounds. In this chapter, we will identify various resources for finding government employment, discuss the types of government jobs available, explain the job application and selection process, and highlight various military-specific issues of relevance to government employment. While the focus is on federal employment, we address state and local employment at the end of this chapter.

Advantages of Government Service

Working in the public sector has many important advantages:

- You have first-hand experience working for the federal government and know the culture and how the government operates. In many cases, there is a direct correlation between the work you performed in the armed forces and the civilian positions.
- Government compensation is competitive with positions in the private sector. This includes health care benefits and a pension plan. See federal pay charts at www.opm.gov/policy-data-oversight/pay-leave/salaries-wages. For those separating short of retirement, your time in service can count towards federal employment.
- Federal employment tends to be more secure than private sector employment.
- Most federal employees are satisfied with their job, finding it interesting and rewarding.
- Federal employment opportunities exist nationwide and even overseas.
- Career progression within the federal government is merit based, resulting in general fairness in promotions and advancement. Your knowledge of the government culture and what it takes to be a top performer gives you a leg up.

Finding the Job

There are numerous federal, military, and commercial resources available to help you find government jobs.

Federal Government

All federal agencies are required to post their jobs, and the USAJobs website (www.usajobs.gov) is often the one selected for that purpose. You can search for jobs by series, agency, location, title, and keywords. There is also a Frequently Asked Questions (FAQ) section.

The website for the Veteran Employment Program Offices (https://www.fedshirevets.gov/veterans-council/agency-directory) lists the agency representatives responsible for promoting veterans' recruitment, employment, training and development, and retention.

You should be aware that the Human Resources departments of all the federal agencies have hardcopies of the federal job announcements available to job seekers.

Military Services

Many of the military services run their own civilian jobs websites. These include the Department of the Navy (www.secnav.navy.mil/donhr), the Department of the Army (www.goarmy.com/careers-and-jobs/army-civilian-careers.html), the Air Force (www.afciviliancareers.com) and the Coast Guard (www.dcms.uscg.mil - search "civilian hr").

Commercial Websites

There are also several commercial websites that list government jobs, including Federal Job Search (www.federaljobsearch.com) and Federal Government Jobs (www.federaljobs.net). Some organizations offer fee-based federal job search assistance, such as Battlefield Resumes (www.battlefieldresumes.com) and the Resume Place (www.resume-place.com).

In addition to the above resources, you are encouraged to talk with friends and associates who are working for the government to learn about job opportunities. Let them know you are job searching and ask them to keep you in mind as opportunities arise.

Understanding Federal Jobs

Each federal job announcement includes the following:

- Job Title
- Series - Jobs are defined by a 3- or 4-digit "series" number similar to a Military Occupational Specialty code or rating. For a list of all the series numbers and job titles, visit www.opm.gov/policy-data-oversight/classification-qualifications/classifying-general-schedule-positions/occupationalhandbook.pdf.
- Grade - Relates directly to the salary for the position.
- Position Information – The type of position (temporary, term, permanent).
- Duty Location – The location of the position.
- Who May Be Considered – Who is eligible to apply for the position.
- Duties – The job duties and responsibilities.
- Specialized Experience – Experience required to perform the job.
- KSAs (Knowledge, Skills, & Abilities) or Qualifications and Evaluations – Requirements that need to appear in your resume.
- How To Apply – This provides specifics on how and where to submit your job application and the closing date for applying to the position.

Competitive vs. Excepted Service

Federal agencies offer their civilian positions on a competitive or on an excepted basis. The Office of Personnel Management's Human Resource office provides oversight for the hiring process in both cases. Most commonly, federal civilian jobs are competitive service positions. These jobs are posted on USAJOBS.gov, and applicants take part in an open competition for them.

When it is not viable for an agency to use the conventional competitive process, OPM provides a hiring authority to allow the agency to streamline the hiring process. This is the case for excepted service jobs, and these jobs are not required to be posted on USAJOBS. As some of these positions are only announced on the agency's website, it is important to look at individual agency websites for job announcements as well. More information on excepted service can be found at https://www.opm.gov/policy-data-oversight/hiring-information/excepted-service.

Applying for the Position

Follow these steps when applying for a federal position:

1. *Carefully review the job announcement.* Are you eligible to apply? Are you qualified to apply? Are you comfortable with the duties, and have you performed them in the past? Do you have the specialized experience and KSAs/competencies the government agency is seeking?

2. *Write your federal resume.* There are multiple ways to apply for a federal job, so follow the job posting directions carefully.
   a) The recommended way to apply is to create your resume using the Resume Builder found in USAJOBS.gov. Tailor your resume to the duties of the position and explain how your experience meets the required specialized experience. Include your country of citizenship, highest grade level if you have previous federal employment, and how much Veteran's preference you have, if any. Also include the hours per week you worked at each position, as this information is mandatory. See the federal resume sample at the end of this chapter for more tips.
   b) You can upload your resume (if you do this include the last four of your social security number on every page).The federal resume, unlike the corporate resume, should be long (4 to 5 pages is average) and detailed.
   c) If you are submitting a hard copy application, place the position title and job announcement number on each page.
   d) You can use the brochure *OF 510 Applying for a Federal Job* for guidance.
   e) Some agencies may require the use of *OF 612 Optional Application for Federal Employment.*

3. *Give careful thought to the KSA/competencies section.* This is the most important part of the federal job application process and is normally listed under "Qualifications and Evaluations." You must incorporate the KSAs/competencies into your resume to show you possess them or you will not qualify for the position. Think of the KSAs as interview questions, and cite 3-4 sentence examples from your work experience to match each.

4. *Complete the package.* Your completed job application package should consist of a cover letter (optional), your resume, your short responses addressing each KSA, and any additional information or optional forms that are specified in the job announcement (forms

found at www.opm.gov). Veterans are usually asked to provide the actual VA letter that states they have a certain percentage of disability. Include this letter in your application. Common mistakes in the job application process include leaving required fields blank, not marking VRA/VEOA (Veterans Employment Opportunity Act), not using the allotted number of characters per experience block, and not following directions. As in the military, attention to detail is important!

The Selection Process

The hiring agency will review your job application for completeness and basic qualifications. Your application resume will be compared against the required qualifications and KSAs/competencies, and a score and numerical rating will be rendered. The applications are then ranked according to their scores. Hiring Reform allows for any application package with a rating of 70 or more points to be referred to the hiring manager for consideration.

Applicants with Veteran's Preference are placed at the top of each category (gold, silver, or bronze), and considered first. The hiring manager reviews each application package and makes the final decision. If you are selected, you will receive an interim offer letter contingent on a background/security check.

Veterans Preference

Veterans discharged or released from active duty in the armed forces under honorable conditions during a time of war are eligible for veteran's preference. If you are a retired member of the armed forces you are not included in the definition of preference eligible unless you are a disabled veteran or you retired below the rank of major or its equivalent. Visit www.fedshirevets.gov for details on Veterans' Preference.

The government hiring authority can name select you - if you are qualified - up to the GS-11 level. The Veteran's Recruitment Appointment (VRA) applies to all active duty military but not to those who have been out of the military for more than two years.

State and Local Government Employment

Nearly 17 million government jobs are found at the state and local levels. The most numerous state and local government employers are found in education (universities and school districts) and criminal justice (police departments and court systems).

One military transition program you may be interested in is **Troops to Teachers** (www.ProudToServeAgain.com). This program assists transitioning service members and veterans in beginning new careers as K-12 school teachers in public, charter, and Bureau of Indian Affairs schools.

Some useful websites for careers in criminal justice and public safety include: www.911hotjobs.com and www.theblueline.com.

The best approach to finding employment with state and local government agencies is to target particular units of government. Each has their own hiring requirements, which usually involve submitting applications and/or resumes and taking any required tests. Many local governmental units will have a website with information on their application procedures. Also, consider attending military-friendly job fairs, which often include representatives from police departments and schools that are eager to hire transitioning military personnel.

For a list of Federal employment resource links and a step-by-step Federal Employment Checklist. visit www.CorporateGray.com > Transition Guide > Gain Federal Employment.

USAJOBS RESUME BUILDER

The recommended way to apply for a federal job is by using the USAJOBS Resume Builder.

Step 1: Sign in to your USAJOBS account at https://www.usajobs.gov.
Step 2: Click the "Documents" tab to locate the USAJOBS Resume Builder.
Step 3: Click the "Upload or build resume" button and select "Build resume" to open the Resume Builder.
Step 4: Fill in the USAJOBS work experience resume builder, as shown here:

Employer Name
Battlefield Resumes

Employer Address 1
10 Hope Road

Employer Address 2 (optional)

Country
United States

Postal Code
22554

City/Town
Stafford

State/Territory/Province
Virginia

Formal Job Title
Intelligence Analyst

Start Date
January

Start Year
2015

End Date
Present

End Year
Present

Salary (optional)

Currency
USD

Salary Rate
Bi-weekly

Average Hours per week
40

May we contact your supervisor?
(●) Yes ( ) No ( ) Contact me first

Supervisor Name
Bruce Benedict

Supervisor Phone
123-456-7891

Is this a Federal Civilian position?
( ) Yes (●) No

Duties, Accomplishments, and Related Skills

(4257 characters remaining)

KSA 1. ABILITY TO ANALYZE/EVALUATE INTELLIGENCE INFORMATION.
I not only evaluated the analytical products, but I was often asked to evaluate the products of other offices within the J2. I used Structured Analytical Techniques for Intelligence analysis, which were all qualitative analytical methodologies. I am intimately familiar with DCGS-A, TIGR, PALANTIR, CPOF, M3, HOTR, CIA WIRE, INTELINK, DAIIS.

KSA 2. ABILITY TO COMMUNICATE ORALLY AND IN WRITING.

Hiring Manager's View

After you complete the USAJOBS Resume Builder and submit your resume, your work experience will look like this to the hiring manager:

Battlefield Resumes
10 Hope Road
Stafford, VA 22554 United States

01/2015 – Present
Hours per week: 40
Intelligence Analyst
Duties, Accomplishments and Related Skills:

KSA 1. ABILITY TO ANALYZE/EVALUATE INTELLIGENCE INFORMATION.
I not only evaluated the analytical products, but I was often asked to evaluate the products of other offices within the J2. I used Structured Analytical Techniques for Intelligence analysis, which were all qualitative analytical methodologies. I am intimately familiar with DCGS-A, TIGR, PALANTIR, CPOF, M3, HOTR, CIA WIRE, INTELINK, DAIIS.

KSA 2. ABILITY TO COMMUNICATE ORALLY AND IN WRITING.
I personally briefed the Commander on several occasions, and J2 and J3 regularly, on matters concerning emergency preparedness and mitigation. I provided weekly briefings with the J3 and daily briefings with the J2 dealing with issues such as force protections and public opinion.

Supervisor: Bruce Benedict (123-456-7891)
Okay to contact this Supervisor: Yes

chapter 11

Start Your Own Business

WHILE THE PREVIOUS CHAPTERS examined strategies for finding employment in other people's organizations, you may also be interested in working for yourself. Indeed, each year nearly seven percent of Americans start their own business. If your self-assessment activities in Chapter 4 indicate your motivated abilities and skills are very entrepreneurial, you may be well-suited for self-employment.

The Military Entrepreneur

While most military veterans choose to work for the government, corporations, or nonprofit organizations, some decide to go into business for themselves. Indeed, approximately 2.52 million businesses are majority-owned by veterans, according to a 2017 report by the U.S. Small Business Administration's Office of Advocacy (www.sba.gov/sites/default/files/advocacy/435-veteran-owned-businesses-report.pdf). This represents 9.1 percent of all businesses in the United States. Veteran-owned businesses had combined receipts of over $1 trillion and employed 5 million people.

Many of the skills and personal qualities you acquired while in the Service, such as patience, drive, perseverance, sacrifice, problem solving, and handling adversity, are well suited for becoming a successful entrepreneur.

The fact that you may have little or no business experience should not dissuade you from considering an entrepreneurial option to employment. If you have the right combination of skills, some great ideas, and the necessary drive to be an entrepreneur, you should consider working for yourself.

Although the majority of business start-ups fail within the first five years, for those that try again, over 80% are successful. Perseverance and resourcefulness are key characteristics for an entrepreneur. There are a wealth of resources offered by government agencies and nonprofit organizations to assist veterans in establishing their own business. Taking advantage of the programs and services mentioned in this chapter can significantly increase the probability of success among veteran entrepreneurs.

Examine Risks and Motivations

You will find few challenges riskier and more time-consuming than starting your own company. It involves a lot of hard work, long hours, and betting on your financial future. At the same time, you may experience your greatest professional satisfaction in running your own business, especially if you are pursuing your passion. You also may achieve considerable wealth. Indeed, small business owners disproportionately earn much more than the average employee.

Starting a business means taking risks. While you may have grandiose visions of becoming an overnight success, you will probably realize little income during the first two years of building your business. It takes time to develop a regular clientele. What profits you do realize are normally plowed back into the business in order to expand operations and realize larger future profits.

At the same time, however, owning your own business can be tremendously satisfying. It is the ultimate exercise of independence and being in control. You are rewarded in direct proportion to your productivity. Your salary is not limited by a boss, nor are your accomplishments credited to others. Depending on how successful your business becomes, you may be able to retire young and pursue other interests. You can turn what you truly enjoy doing into a profitable, rewarding, and fun career.

Possess the Right Strengths for Success

How can you become self-employed and successful at the same time? No one has a magical success formula for the budding entrepreneur—only advice based on experience. We do know why many businesses fail, and we can identify some basic characteristics for success.

Poor planning, management, and decision-making lie at the heart of business failures. Many people go into business without doing sufficient market research. They undercapitalize, select a poor location, incur extremely high overhead costs, lack a commitment, or lack interpersonal and salesmanship skills.

Successful business people combine certain motivations, skills, and circumstances. Contrary to popular myths, you don't need to be rich or have an MBA or business experience to get started. If you are willing to gamble and are a self-starter, self-confident, an organizer, and you like people, you should consider this entrepreneurial alternative in your career transition plan. These characteristics, along with drive, thinking ability, communication and networking skills, technical knowledge, hard work, persistence, good luck, and a good idea to fill a need are essential ingredients for business success.

If these are among your strengths, as identified in Chapter 4, you may be a good candidate for starting your own business with a high probability of success. If you feel you have recurring weaknesses in certain areas, you may want to consider finding a business partner who has particular complementary strengths for running a business.

Franchising

According to the International Franchise Association's VetFran website (www.vetfran.org), one out of every seven franchises in the United States is owned by a veteran. Although veterans account for approximately 7% of the total population, 14% of franchisees are vets. Military veterans can make successful franchise owners namely because they typically have a strong work ethic and the ability to follow the rules, key attributes for a franchisee. The VetFran Program, or Veterans Transition Franchise Initiative, helps transitioning service members access franchise opportunities through training, financial assistance, and industry support.

There are literally thousands of franchises to choose from, and over 630 companies offer discounts to prospective vet franchisees. Some require a relatively small investment and a modest amount of operating capital; others cost in excess of a million dollars. Some franchises are service-based, while others are product based.

You can get individualized assistance in selecting a franchise from FranNet.com (https://frannet.com). Several websites include directories of franchise opportunities, such as Entrepreneur.com's website at www.entrepreneur.com/franchise500, www.franchisesolutions.com, and the International Franchise Association's website at www.franchise.org. These sites allow you to search for franchises by investment, industry and location.

Getting Started

You should do two things before you decide to go into business. First, thoroughly explore your goals and motivations. The questions are familiar:

- What do you want to do?
- What do you do well?
- What do you enjoy doing?

Second, research different types of businesses in order to better understand advantages, disadvantages, procedures, processes, and possible problems. Talk to business owners about their work. Contact fellow veterans who have started their own businesses, such as those affiliated with Veterans Business Services (veteransbusinessservices.us) or found through directories of veteran-owned businesses such as Veteran Owned Business (veteranownedbusiness.com) and National Veteran-Owned Business Association (navoba.org).

Try to learn as much as possible about the reality before you invest your time and money. Surprisingly, few people do this. Many people leap into a business that they think will be great and then later learn it was neither right for them nor did they have realistic expectations of what was involved.

You should approach business opportunities the same way you approach the job market: do research, develop networks, and conduct informational and referral interviews. Most business people, including your competition, will share their experiences with you and assist you with advice and referrals. Such research is invaluable. Don't be high on motivation but low on knowledge and skills, for "thinking big" is no substitute for doing the hard work involved in launching and operating the many details involved in running a successful business.

Look for New Opportunities

Most business people will tell you similar stories of the reality of running your own business. Do your market research, work long hours, plan, and be persistent. They also will give you advice on what businesses to avoid and what business routines you should be prepared to handle.

Many service and high-tech businesses will continue to grow in the decade ahead. Given major issues, such as energy, the environment, health care, fitness, and leisure, and the changing demographic structure—fewer young people, more elderly, the two-career family—businesses relating to home maintenance, recreation, health care, housing for the elderly, green technologies, energy, the environment, and mortuaries and cemeteries should expand considerably during the next two decades.

Opportunities are also available for inventive business people who can make more productive use of busy people's time—for example, fast food, financial planning, and online shopping. The information and high-tech revolutions are taking place at the same time two-career families do not have time to spend standing in lines at banks, grocery stores, and department stores.

A service business is especially attractive. It is easy to establish; many require a small initial investment; and the accounting is often simple. You may be able to operate from your home and thus keep overhead down.

Knowing these trends and opportunities is important, but they should not be the only determining factors in choosing a business. You should start with yourself by again trying to identify a business that is fit for you rather than one you think you might fit into.

Prepare the Basics

You need to consider several other factors before starting a business. Since a business requires financing, locating, planning, developing customer relationships, and meeting legal requirements, be prepared to address these questions:

1. **How can I best finance the business?** You could bootstrap with your own savings, take out a personal or business loan with a bank, go into a partnership to share the risks and costs, or explore special loan programs for veterans from the Small Business Administration (www.sba.gov).
2. **How much financing do I need?** Many businesses fail because they are under-capitalized. Others fail from over-spending on rent, inventory, personnel, and advertising.
3. **Where is my market?** Just in this community, region, nationwide, or international?
4. **Who are my suppliers?** How many must I work with? What about credit arrangements and payment terms? Are some suppliers willing to ship directly to my customers?
5. **Where is the best location for the business?** Should I open a store, operate from my home, or conduct a virtual (Internet) business?
6. **How should the business be legally structured?** Sole proprietorship, partnership, or corporation? Each has certain advantages and disadvantages.
7. **What licenses and permits do I need?** These consist of local business licenses and permits, federal employee identification numbers, state sales tax number, state occupational licenses, and federal licenses and permits.
8. **What type of insurance do I need?**
9. **How many employees do I need and how do I find them?** Can I do without personnel initially until my business expands? Should I use part-time and temporary help? How should I recruit (Internet, classified ads, employment firms, headhunters, or network)?
10. **What business name should I use?** If incorporated, is anyone else using the name? If a trade name, is it registered?
11. **What accounting system should I use?** Cash or accrual? Can I handle the books or do I need an accountant? Who will handle the timely calculation and payment of payroll, sales, and corporate taxes, insurance, retirement plans, and workers compensation?
12. **What technology and communications decisions do I need to make?** What type of website should I develop? What types of computer and telecommunication equipment should I acquire? Is it best to buy or lease business equipment?
13. **Do I need a lawyer?** What type of lawyer? What legal work can I do myself?
14. **How do I develop a business plan?** A business plan should include a definition of the business, a marketing strategy, operational policies, purchasing plans, financial statements, and capital-raising plans.
15. **Should I put together a board of advisors?** It's always good to include respected individuals who can contribute expertise and perspective to your operation.

Get Useful Advice

Research and get useful advice before selecting and starting your own business. You will find a wealth of information from the **Small Business Administration's (SBA) Office of Veterans Business Development** (www.sba.gov/about-sba/sba-locations/headquarters-offices/office-veterans-business-development). Also visit the SBA's Boots to Buisness website which includes courses for veterans on starting a business: https://sbavets.force.com/s.

Several other websites can provide assistance in starting a business:

- Entrepreneur.com https://www.entrepreneur.com/topic/starting-a-business
- Business Know-How www.businessknowhow.com
- AllBusiness www.allbusiness.com

The federal government will help you with several publications available through the SBA. The Internal Revenue Service (www.irs.gov/businesses/small-businesses-self-employed/starting-a-business) offers a great deal of tax and accounting information to start-up businesses. The U.S. Chamber of Commerce (www.uschamber.com) as well as its local chapters offer useful information. Other good online resources to help you start a business include:

- Business Owner's Toolkit www.bizfilings.com/toolkit
- CEO Business Express www.ceoexpress.com
- Entrepreneurship.org www.entrepreneurship.org

Acquire Veterans Assistance

Several government agencies and nonprofit organizations offer assistance to veterans interested in starting their own business. Most of the assistance involves training, counseling, and financing. Be sure to explore the following websites and programs designed specifically for entrepreneurial veterans:

- Veterans Business Services www.veteransbusinessservices.us
- SBA Veterans Business Outreach Center (VBOC) www.sba.gov/page/veterans-business-outreach-center-vboc-program
- Department of Veterans Affairs OSDBU www.va.gov/osdbu
- VetFriends www.vetfriends.com
- DoL Veterans Employment and Training Center www.dol.gov/agencies/vets
- Institute for Veterans & Military Families (IVMF) https://ivmf.syracuse.edu/programs/entrepreneurship

Government Set-Asides and Sole-Source Contracts

If you receive a disability rating by the Department of Veterans Affairs upon leaving the Service, you can apply to have your business certified as a Service Disabled Veteran Owned Small Business (SDVOSB). There are significant benefits to having this designation. For more information, visit the SBA website at www.sba.gov/federal-contracting/contracting-assistance-programs/service-disabled-veteran-owned-small-businesses-program#.

For more information about starting your own business, visit www.CorporateGray.com > Transition Guide > Start Your Own Business.

MAKING THE MOST OF A JOB FAIR

Attending a job fair, whether in-person or virtual, can be one of the most effective job search tools you employ. Here are some ideas to help you make the most of the next job fair you attend.

Proactively Prepare

Be proactive in preparing for the event. Research the attending companies and apply to positions of interest before the event. Also prepare questions about the companies and relevant information about yourself to share with the recruiters you will meet. This will help you take full advantage of the one-on-one interactions that the job fair offers.

Use LinkedIn before the event to identify your connections at participating companies. Solicit their feedback and seek their assistance in identifying appropriate positions. Apply in advance to those jobs, and then reference those specific opportunities at the job fair, letting the recruiter know how you are a match.

Make the Most of the Event

Use your research of the companies to go beyond your resume by demonstrating your enthusiasm and interest in the company that cannot be conveyed on paper. Make the most of your short time with each company representative. Share relevant and interesting information about yourself to be memorable to the recruiter.

Ask the questions your prepared as you researched the companies. Get as much information as you can from the recruiters and their brochures. Learn about the company's needs and culture to see if it is a fit for you. You are screening the recruiter just as they are screening you. Ask the recruiters for their contact information so you can follow-up with them later.

Follow Up

Review your notes after the event and give careful thought to your next moves. Keep a record of the contacts you made and the next steps you should take in the job search process.

Contact those you met at the event to thank them for the information they provided. Reiterate why you are match for their company's opportunities. If you engage with recruiters again in a follow-up email or via LinkedIn, you'll increase the chances of making a positive impression and being memorable for future opportunities.

Corporate Gray has been a long-time leader in the job fair industry. We believe there is value in meeting with company representatives, getting answers to your questions, and making a positive first impression.

See the list of Corporate Gray Military-Friendly Job Fairs on the next page, and visit www.CorporateGray.com/jobfairs to learn more and register today.